COMPUTATIONAL
METHODS IN
GEOSCIENCES

COMPUTATIONAL

METHODS IN

GEOSCIENCES

Edited by W. E. Fitzgibbon
University of Houston

Mary Fanett Wheeler
Rice University

siam.

Philadelphia

Society for Industrial and Applied Mathematics

COMPUTATIONAL METHODS
IN GEOSCIENCES

Library of Congress Cataloging-in-Publication Data

Computational methods in geosciences / edited by W. E. Fitzgibbon, Mary
   Fanett Wheeler.
        p.   cm.
    Papers based on a conference held in Sept. 1989 and hosted by the
Society for Industrial and Applied Mathematics with the cooperation
of the Dept. of Mathematics and the Energy Laboratory of the
University of Houston.
    Includes bibliographical references.
    ISBN 0-89871-301-3
    1. Earth sciences—Mathematics.   I. Fitzgibbon, W. E. (William
Edward), 1945–   II. Wheeler, Mary F. (Mary Fanett)
III. Society for Industrial and Applied Mathematics.
QE33.2.M3C65   1992
550'.15'1—dc20   92-16402

PREFACE

In September of 1989, the Society for Industrial and Applied Mathematics, with the cooperation of the Department of Mathematics and the Energy Laboratory of the University of Houston, hosted a conference on mathematical and computational issues in geophysical fluid and solid mechanics. This was the third in an ongoing sequence of SIAM conferences pertaining to the geosciences.

The purpose of this conference was to provide a forum where mathematicians, geophysicists, geologists, hydrologists, and petroleum engineers could meet, discuss, and collaborate on problems of mutual interest. The central topics were systems of conservation laws, reactive flows, fluid and solid mechanics, partial differential equations, wave propagation, materials response, and geochemistry. Areas of application include flow in porous reservoirs and acquifers, basin modeling, seismic modeling and inversion contaminant transport, and remote sensing.

The program of the conference consisted of invited plenary lectures, minisymposia or invited special sessions, research workshop contributed papers, poster sessions, and informal seminars and discussions.

These volumes are not intended to be a proceedings of the conference per se. Participants in the conference were encouraged to submit manuscripts developed from the topics presented at the meeting. The submissions could either be expository papers or original research. The standard elongated abstract was not acceptable. All papers were refereed by outside reviewers.

The accepted papers fall roughly into three categories which we have somewhat arbitrarily labeled as Computational Methods in Geosciences, Modeling and Analysis of Diffusive and Advective Processes in Geosciences, and Wave Propagation and Inversion. These broad categories serve as the titles of this sequence of three volumes. We realize that our categorization is imperfect and hope that it does not serve to confuse or offend.

If dedications are appropriate, these volumes should be dedicated to Garrett T. Etgen, Chairman of the Mathematics Department at the University of Houston, who was tremendously supportive of the endeavor, worked endlessly, and derived no credit or visibility. In this dedication we show our deepest appreciation for his tireless efforts to further the development of mathematical sciences in the Houston area.

The errors that most assuredly occur in these volumes are consequences of the carelessness and incompetence of the editors and the editors herewith and henceforth apologize.

W. E. Fitzgibbon
Mary Fanett Wheeler

CONTENTS

CHAPTER 1

The Pseudospectral Method for Limited-Area Elastic Wave Calculations*

Jeffrey M. Augenbaum†

Abstract. We introduce a Chebyshev based pseudospectral method for elastic wave propagation suitable for limited area modeling. The method is based on the velocity-stress first order system. Finite differences are used for the time integration; and spectral differentiation for the spatial derivatives. Boundary conditions based on characteristic variables allow a stable treatment of the free surface boundary and artificial (absorbing) boundaries. The method is compared with a fourth order finite difference scheme of Bayliss et al. [3] for the Lamb problem and for scattering from a plane interface. We find that the spectral method requires approximately one-half the resolution in each coordinate direction as compared to the fourth order finite difference scheme to give comparable accuracy.

Key words. elastic, wave propagation, pseudospectral

1. Introduction.

Accurate numerical simulations of elastic waves have become an increasingly important and cost effective tool for the study of wave propagation through realistic elastic media, where exact, or other, methods are not available. Most often, low order finite difference methods, because of their simplicity, or finite element methods, because of their ability to handle complex geometry, have been used for the spatial discretization. Both of these methods are limited by the need to use 10 to 20 grid points per wavelength to accurately resolve the propagating waves [7]. Typical frequencies encountered in realistic seismic problems suggest that the cost of 2D or 3D calculations can be prohibitive, even with present day supercomputers.

Spectral methods [4,5,10,18] have become the numerical method of choice for the approximate solution of time dependent partial differential equations in many branches of computational science. They consist of expanding the solution of a differential equation in terms of a series of global, infinitely differentiable, basis functions. An efficient implementation, especially well suited to nonlinear equations and equations with spatially varying

---

* revised May 1990
† Department of Mathematics, U-9 University of Connecticut, Storrs, CT 06269

coefficients, is the *spectral collocation technique* also known as the *pseudospectral method*. In this approach the fundamental unknowns are the solution values at a set of collocation points and the series is used solely for the purpose of approximating derivatives. This can be accomplished efficiently for certain polynomial basis functions by fast transform methods. As a result of the global nature of the approximation, spectral methods are able to yield highly accurate approximations of smooth solutions with substantially fewer grid points than would be required by local (finite difference or finite element) methods. Whereas finite difference and finite element discretizations give algebraic convergence i.e. the discretization error is $O(N^{-p})$, where N is the number of grid points or local basis functions, and p is typically a small integer (usually 2 or 4), spectral methods have been shown (for analytic solutions) to have exponential convergence $O(e^{-\alpha N})$ for some positive number α.

Spectral methods were first used for forward acoustic modeling in [9,11], and for elastic modeling in [12]. Further studies, comparing the efficiency and accuracy of pseudospectral and finite difference methods, are contained in [6,8,7,16]. The number of grid points required by spectral methods to solve a model wave equation with an accuracy comparable to that of a fourth order finite difference method was shown [7] to yield a reduction in the number of grid points by a factor of 4. (The savings in computer resources can be quite substantial in 3D [14,15].) In order to use a uniform distribution of collocation points (similar to finite difference methods), all these studies use trigonometric basis functions; and hence must assume periodic boundary conditions. This seems to be an unnecessary restriction as it forces one to use needlessly large computational domains in order to avoid the unphysical wave interactions that result through the use of periodic boundary conditions. Realistic boundary conditions such as free surfaces and sloping internal layers are treated in an unnatural way [11,12]. Moreover, it precludes the use of domain decomposition methods [1,2] to treat problems in layered media, as this leads to nonperiodic boundary conditions on the subdomains.

Nonperiodic, and more realistic, boundary conditions can be enforced by using other orthogonal basis functions. The Chebyshev polynomials, for example, are commonly used in fluid dynamic calculations [4,5]. They posses the faster than algebraic convergence rates typical of the spectral method, and differentiation can still be done efficiently by using fast transform techniques. A Chebyshev based spectral method has been tested for forward acoustic modeling with simplified, dirichlet and neumann, boundary conditions in [13].

The purpose of this paper is to introduce a Chebyshev based pseudospectral method for elastic wave propagation suitable for limited area modeling. The method is based on the velocity-stress first order system. Finite differences are used for the time integration; and spectral differentiation for the spatial derivatives. Boundary conditions based on characteristic variables allow a stable treatment of the free surface boundary and artificial (absorbing) boundaries.

The paper is organized as follows. The elastic equations along with some basic properties are given in the section 2, followed by a detailed discussion of the numerical scheme in section 3. Numerical examples comparing the spectral method with a fourth order finite difference scheme for the Lamb problem and elastic scattering from a horizontal interface are given in section 4. Concluding remarks are then presented in section 5.

2. Elastic Wave Equation.

Basic Equations.
The elastic equations are most commonly written as a system of second order wave equations for the displacement vector, $\mathbf{U}(\mathbf{x}, t)$, of a material particle from its equilibrium

position $\mathbf{x} \in R^3$. In two dimensions they take the form

(2.1)
$$\begin{aligned}
\rho U_{tt} &= [\lambda(U_x + V_y) + 2\mu U_x]_x + [\mu(U_y + V_x)]_y + \rho f_1 \\
\rho V_{tt} &= [\mu(U_y + V_x)]_x + [\lambda(U_x + V_y) + 2\mu V_y]_y + \rho f_2
\end{aligned}$$

Here the Lamé parameters $\lambda = \lambda(\mathbf{x})$ and $\mu = \mu(\mathbf{x})$ as well as the density $\rho = \rho(\mathbf{x})$ can vary in space; and $\mathbf{x} = (x, y) \in \Omega \subset R^2$, where Ω is the domain of interest. The external body forces are denoted by the source functions $f_1(\mathbf{x}, t), f_2(\mathbf{x}, t)$.

Alternatively, if we uses the stress-strain relations

(2.2)
$$\begin{aligned}
\tau_{11} &= \lambda(U_x + V_y) + 2\mu U_x \\
\tau_{22} &= \lambda(U_x + V_y) + 2\mu V_y \\
\tau_{12} &= \mu(U_y + V_x)
\end{aligned}$$

we can derive a first order system of equations for the velocity (u, v) and stress $(\tau_{11}, \tau_{22}, \tau_{12})$ terms

(2.3)
$$\begin{aligned}
\rho u_t &= \tau_{11,x} + \tau_{12,y} + \rho f_1 \\
\rho v_t &= \tau_{12,x} + \tau_{22,y} + \rho f_2 \\
\tau_{11,t} &= (\lambda + 2\mu)u_x + \lambda v_y \\
\tau_{22,t} &= \lambda u_x + (\lambda + 2\mu)v_y \\
\tau_{12,t} &= \mu(v_x + u_y)
\end{aligned}$$

where $(u, v) = (U_t, V_t)$ represents the velocity of the material particles.

We can also write (2.3) in vector form

(2.4)
$$W_t = AW_x + BW_y + F$$

where $W = (u, v, \tau_{11}, \tau_{22}, \tau_{12})^T$ is the solution vector, $F = (f_1, f_2, 0, 0, 0)$ is the source vector, and the matrices A and B are given by

(2.5)
$$A = \begin{pmatrix}
0 & 0 & 1/\rho & 0 & 0 \\
0 & 0 & 0 & 0 & 1/\rho \\
\lambda + 2\mu & 0 & 0 & 0 & 0 \\
\lambda & 0 & 0 & 0 & 0 \\
0 & \mu & 0 & 0 & 0
\end{pmatrix}$$

(2.6)
$$B = \begin{pmatrix}
0 & 0 & 0 & 0 & 1/\rho \\
0 & 0 & 0 & 1/\rho & 0 \\
0 & \lambda & 0 & 0 & 0 \\
0 & \lambda + 2\mu & 0 & 0 & 0 \\
\mu & 0 & 0 & 0 & 0
\end{pmatrix}$$

Dispersion Relations.

We seek homogeneous solutions of (2.4) (i.e. F = 0) of the form

(2.7)
$$W(x, y, t^n) = e^{i(\xi x + \eta y)} \hat{W}(t^n)$$

Substituting (2.7) into (2.4) gives

(2.8)
$$\hat{W}_t + G\hat{W} = 0$$

where

(2.9)
$$G = i(\xi A + \eta B)$$

whose solution is

(2.10)
$$\hat{W}(t) = e^{-Gt}\hat{W}(0)$$

The nonzero eigenvalues of G are pure imaginary. They can be written as $i\omega(\xi, \eta)$ where

(2.11)
$$(\omega_{1,2})^2 = (\xi^2 + \eta^2)(\frac{\lambda + 2\mu}{\rho})$$

(2.12)
$$(\omega_{3,4})^2 = (\xi^2 + \eta^2)(\frac{\mu}{\rho})$$

(2.13)
$$\omega_5 = 0$$

The nonzero eigenvalues represent the frequencies of the motion. They correspond to a propagating compression (P) wave and shear (S) wave with speeds

(2.14)
$$c_p^2 = \frac{\lambda + 2\mu}{\rho}$$
$$c_s^2 = \frac{\mu}{\rho}$$

The zero eigenvalue corresponds to a nonpropagating mode and can be thought of as a compatibility condition for the stresses.

3. Numerical Method.

Spatial Discretization.

We now briefly describe the Chebyshev spectral collocation method for two dimensional problems.

Consider a function $u(x, y)$ on the computational domain $(x, y) \in [-1, 1] \times [-1, 1]$. We can approximate $u(x, y)$, based on its values at the set of Chebyshev (Gauss-Lobatto) collocation points

(3.1)
$$x_i = \cos\frac{\pi i}{M} \qquad i = 0, \ldots, M$$
$$y_j = \cos\frac{\pi j}{N} \qquad j = 0, \ldots, N$$

,by the MN-th degree Chebyshev interpolating polynomial

(3.2)
$$\mathcal{P}_{MN}u(x, y) = \sum_{i=0}^{M}\sum_{j=0}^{N} u(x_i, y_j)\psi_{M,i}(x)\psi_{N,j}(y)$$

with

(3.3)
$$\psi_{n,i}(x) = \frac{(-1)^{i+1}(1 - x^2)T_n'(x)}{\bar{c}_{n,i}n^2(x - x_i)}$$

where

$$T_n(x) = \cos(n\cos^{-1}x)$$

and $\bar{c}_{n,0} = \bar{c}_{n,n} = 2$, and $\bar{c}_{n,j} = 1$, for $1 \le j \le n - 1$.

Note that since $\psi_{M,i}(x_k) = \delta_{ik}$, and $\psi_{N,j}(y_k) = \delta_{jk}$ (where $\delta_{ik} = 1$ for $i = k$, and $=0$ for $i \ne k$) (3.2) is exact at the collocation points.

Differentiation is then accomplished by differentiating (3.2) at the collocation points. We therefore get, [4,5,18]

$$(3.4) \qquad \left(\frac{\partial u}{\partial x}\right)_{x=x_k} = \sum_{i=0}^{M} \sum_{j=0}^{N} u(x_i, y_j)\psi'_{M,i}(x_k)\psi_{N,j}(y) = \mathbf{D}_M \mathbf{U}$$

and

$$(3.5) \qquad \left(\frac{\partial u}{\partial y}\right)_{y=y_l} = \sum_{i=0}^{M} \sum_{j=0}^{N} u(x_i, y_j)\psi_{M,i}(x)\psi'_{N,j}(y_l) = \mathbf{U}(\mathbf{D}_N)^T$$

where

$$\mathbf{D}_M = \begin{cases} D_{00} = \frac{2M^2+1}{6} & \\ D_{kj} = \frac{\bar{c}_{M,k}(-1)^{j+1}}{\bar{c}_{M,j}(x_k - x_j)} & k \ne j \\ D_{jj} = \frac{x_j}{2(1-x_j^2)} & j \ne 0, M \\ D_{MM} = -D_{00} & \end{cases}$$

is the $(M + 1) \times (M + 1)$ Chebyshev differentiation matrix and $\mathbf{U} = u_{ij} = u(x_i, y_j)$ is an $(M+1) \times (N+1)$ matrix containing the values of u at the collocation points (x_i, y_j). These matrix matrix multiplications can be performed in $O(NM^2)$ (or $O(MN^2)$) operations, however, this can be very efficient on a vector or parallel machine [17].

Alternatively (3.2) can also be written as

$$(3.6) \qquad \mathcal{P}_{MN}u(x,y) = \sum_{m=0}^{M} \sum_{n=0}^{N} a_{mn}T_m(x)T_n(y)$$

where

$$(3.7) \qquad a_{mn} = \frac{4}{MN\bar{c}_{M,m}\bar{c}_{N,n}} \sum_{i=0}^{M} \sum_{j=0}^{N} \frac{u(x_i,y_j)}{\bar{c}_{M,i}\bar{c}_{N,j}} \cos\frac{\pi i}{M} \cos\frac{\pi j}{N}$$

Thus $\mathcal{P}_{MN}u(x_i, y_j)$ can also be evaluated by two 2D FFT's in $O(MNlogMN)$ operations. Differentiation is then accomplished by

$$\frac{\partial u}{\partial x} = \sum_{m=0}^{M} \sum_{n=0}^{N} a_{mn}T'_m(x)T_n(y)$$

$$(3.8) \qquad = \sum_{m=0}^{M} \sum_{n=0}^{N} b_{mn}^{(10)}T_m(x)T_n(y)$$

and

$$\frac{\partial u}{\partial y} = \sum_{m=0}^{M} \sum_{n=0}^{N} a_{mn}T_m(x)T'_n(y)$$

$$(3.9) \qquad = \sum_{m=0}^{M} \sum_{n=0}^{N} b_{mn}^{(01)}T_m(x)T_n(y)$$

where

$$b_{M,n}^{(10)} = 0, \qquad b_{M-1,n}^{(10)} = 2M a_{M,n}$$

$$b_{m,N}^{(01)} = 0, \quad b_{m,N-1}^{(01)} = 2N a_{m,N}$$

and

$$\bar{c}_{m,n} b_{m,n}^{(10)} = b_{m+2,n}^{(10)} + 2(m+1)a_{m+1,n} \qquad 0 \le m \le M - 2$$

$$\bar{c}_{m,n} b_{m,n}^{(01)} = b_{m,n+2}^{(01)} + 2(n+1)a_{m,n+1} \qquad 0 \le n \le N - 2$$

Thus differentiation can also be done by two 2D FFT's. Note that both of the above differentiation methods (matrix multiplication or FFT) yield the derivatives at all collocation points, including the boundary points. Thus the spectral accuracy of differentiation is maintained up to and including the boundary.

Temporal Discretization.

Returning now to the elastic equations (2.4), we approximate the solution at the collocation points (x_i, y_j) by $\mathbf{W}_{ij}(t) = (u_{ij}(t), v_{ij}(t), \tau_{11,ij}(t), \tau_{22,ij}(t), \tau_{12,ij}(t))^T$. Denote the spatially discretized right hand side of (2.4) by G_{ij}^C. Then the elastic system (2.4) can be written as a $(M+1) \times (N+1)$ matrix ODE

$$(3.10) \qquad \frac{d\mathbf{W}_{ij}}{dt} = \mathbf{G}_{ij}^C \qquad i = 0, \ldots, M \quad j = 0, \ldots, N$$

We now integrate equation (3.10) forward in time with any stable ODE solver (Method of Lines) and impose the appropriate boundary conditions after each time step. In our experiments we have used second and fourth order Runge-Kutta schemes.

Timestep and Grid Spacing Requirements.

Use of the Gauss-Lobatto collocation points (3.1) leads to a nonuniform distribution of grid points. The effective resolution is $O(N^{-1})$ near the center of the domain and $O(N^{-2})$ near the boundary. This higher resolution near the boundary is necessary to counteract the tendency of the Chebyshev polynomials to oscillate more severely at the boundaries. For smooth functions however, the pointwise approximation error is essentially uniformly distributed throughout the domain (see [17] for a detailed analysis). The minimum resolution necessary to resolve a wave to an error of at most 10% is π points per wavelength ([10]). This is in contrast to fourier collocation (for periodic functions) where only 2 points per wavelength are required.

The choice of time step involves two considerations, stability and accuracy. In general, explicit finite difference time differencing coupled with Chebyshev spatial differencing has a stability condition of the form

$$(3.11) \qquad \Delta t = O(1/cN^2)$$

Thus for problems with multiple wave speeds, the time step is restricted by the fastest waves. In the elastic case this will be the compressional wave speed (i.e. $\Delta t = O(1/c_p N^2)$). On the other hand the accuracy requirement is determined by the slowest wave speed. That is to say, the number of collocation points, N, is chosen to resolve the smallest wavelengths. Let f_{peak} be the peak frequency that we wish to resolve, then the smallest shear wavelength is given by $\lambda_s = c_s/f_{peak}$ and the smallest compressional wavelength by $\lambda_p = c_p/f_{peak}$. Since $c_s < c_p$ we have $\lambda_s < \lambda_p$. If N_s is the number of collocation points required to resolve λ_s, then the time step restriction becomes

$$(3.12) \qquad \Delta t = O\left(\frac{1}{c_p N_s^2}\right)$$

Note that if the time step were chosen purely on the basis of accuracy, we would choose $\Delta t_{accuracy} = \frac{K}{c_s N_s^2}$. However the actual time step used (3.12) is reduced by a factor $\left(\frac{c_s}{c_p}\right)$, i.e.

$$(3.13) \qquad \Delta t = \frac{K}{c_p N_s^2} = \left(\frac{c_s}{c_p}\right) \Delta t_{accuracy}$$

Boundary Conditions.

The global nature of spectral methods makes them far more sensitive to the boundary treatment than either finite difference or finite element methods. We are interested here in two types of boundary conditions, free surface boundaries and artificial boundaries. In the free surface condition the normal components of stress vanish (i.e. $\tau_{22} = \tau_{12} = 0$ for an upper boundary free surface). Artificial boundaries arise because of truncation of unbounded domains. Here we want to satisfy the equations as best as possible while minimizing spurious reflections caused by the artificial boundary.

The absorbing boundary conditions used here are based on a one dimensional analysis. They are designed to be exact for P and S waves impinging normally on the boundary. Since we are using explicit time integration, the physical variables $\mathbf{W}$ are updated at the boundaries along with the interior points. We then set the linear combination of characteristic variables corresponding to incoming variables to zero and retain only the linear combination corresponding to outgoing characteristic variables. The remaining variable is then updated by extrapolating the linear combination of characteristic variables corresponding to the nonpropagating mode.

At the lateral boundaries ($x = \mp 1$) the characteristic variables are

$$(3.14) \qquad \begin{aligned} R_1 &= \tau_{11} + \rho c_p u \\ R_2 &= \tau_{11} - \rho c_p u \\ R_3 &= \tau_{12} + \rho c_s v \\ R_4 &= \tau_{12} - \rho c_s v \\ R_5 &= \lambda \tau_{11} - (\lambda + 2\mu) \tau_{22} \end{aligned}$$

Similarly, at the top and bottom boundaries ($y = \mp 1$) the characteristic variables are

$$(3.15) \qquad \begin{aligned} R_1 &= \tau_{22} + \rho c_p v \\ R_2 &= \tau_{22} - \rho c_p v \\ R_3 &= \tau_{12} + \rho c_s u \\ R_4 &= \tau_{12} - \rho c_s u \\ R_5 &= (\lambda + 2\mu) \tau_{11} - \lambda \tau_{22} \end{aligned}$$

At the left and top boundaries, (x or y = -1), R_1 and R_3 are the outflow variables while R_2 and R_4 are the inflow variables. At the right and bottom boundaries, (x or y = +1), R_2 and R_4 are the outflow variables while R_1 and R_3 are the inflow variables. In both cases R_5 represents a nonpropagating quantity.

After updating the basic variables, at all collocation points, from $\mathbf{W}^n$ to $\mathbf{W}^{n+1}$ with the temporal scheme (3.10), we denote the basic variables by $\mathbf{W}^*$ and the characteristic variables by $R_j(\mathbf{W}^*)$. We then modify the boundary variables by setting the two outgoing variables equal to the corresponding $R_j(\mathbf{W}^*)$ and the two incoming variables to zero i.e.

$$(3.16) \qquad \begin{aligned} R_{outgoing}(\mathbf{W}^{n+1}) &= R_{outgoing}(\mathbf{W}^*) \\ R_{incoming}(\mathbf{W}^{n+1}) &= 0 \end{aligned}$$

This leads to a 4×4 system of linear equations for $\mathbf{W}^{n+1}$.

We then modify the remaining variable by solving

$$(3.17) \qquad\qquad R_5(\mathbf{W}^{n+1}) = R_5(\mathbf{W}^*)$$

At a free surface we follow a similar procedure. The new outgoing characteristic variables are set equal to the outgoing characteristic variables as computed by the numerical scheme. However, instead of setting the incoming characteristic variable to zero, we use the free surface condition. That is

$$
\begin{aligned}
\tau_{22}^{n+1} &= 0 \\
\tau_{12}^{n+1} &= 0 \\
R_{outgoing}(\mathbf{W}^{n+1}) &= R_{outgoing}(\mathbf{W}^*) \\
R_5(\mathbf{W}^{n+1}) &= R_5(\mathbf{W}^*)
\end{aligned}
$$

(3.18)

This linear system is easily solved for the 3 remaining variables u^{n+1}, v^{n+1}, and τ_{11}^{n+1}. The free surface boundary condition (3.18) can easily be modified to handle a surface line source. For example, we can use $\tau_{22}^{n+1} = S(t^{n+1})\delta(x - x_0)$ in (3.18).

We note that for both types of boundary conditions we have found it necessary to enforce the compatibility condition (3.17) to prevent the onset of spurious, incompatible strain modes (as first reported in [12]). These incompatible strain modes caused the numerical scheme to quickly become unstable.

Algorithm for Chebyshev Pseudospectral Method.

We can now summarize the algorithm for the pseudospectral method for the elastic system (2.3).

(i) Compute the pointwise spatial derivatives of the velocity and stress terms needed for the right hand side of (2.3) at the Gauss-Lobatto collocation points (3.1). Either explicit matrix multiplication or an FFT can be used for this (as described above).

(ii) Evaluate the material parameters ρ, λ, μ at the Gauss-Lobatto collocation points.

(iii) Form the right hand side of (2.3) by pointwise multiplication of the material parameters and the derivatives of the stress and velocity terms.

(iv) Compute the forcing terms due to internal sources (if present) at the interior collocation points.

(v) Update the velocity and stress terms in (2.3) from time t to $t + \Delta t$ at *all* collocation points (including the boundary points). An ODE solver may be used for this.

(vi) Modify the resulting boundary terms to enforce the boundary conditions (as described above).

4. Numerical Results.

In this section we validate the Chebyshev pseudospectral method on two model problems; the Lamb problem for elastic wave propagation in a homogeneous media excited by a surface source and elastic scattering off a plane interface. The results of both of these examples are compared with results generated by the temporally second order and spatially fourth order finite difference scheme of Bayliss et al. described in [3]. The 2-4 finite difference scheme of Bayliss et al. has been previously validated and we therefore consider high resolution simulations (ie. greater than 20 point per wavelength) to be the *exact* solution for purposes of comparison. The spectral calculations presented here all use the second order Runge Kutta method for the time stepping.

Initial Conditions.

In the experiments described below the initial conditions are zero and the model is driven by a source term.

Source Term.

The source function used here is of the form

$$(4.1) \qquad f(x, y, t) = \delta(x - x_0)\delta(y - y_0)S^{(p)}(t)$$

where $\delta(x)$ is a delta function and $S^{(p)}(t)$, the p-th derivative of $S(t)$, gives the time history of the source. In problems with a buried source we use the second derivative of S(t); and for a surface source we use the first derivative of S(t) (in this case the free surface boundary condition become $\tau_{22} = \delta(x - x_0)S^{(1)}(t), \tau_{12} = 0$ in (3.18)).

The discrete delta function is approximated by

$$(4.2) \qquad \delta(x - x_0) = \begin{cases} 0 & x \neq x_0 \\ 1/h_x & x = x_0 \end{cases}$$

where the nominal spatial mesh width is taken to be $h_x = 2/N_x$, and N_x is the number of points in the x direction. The time history $S(t)$ is the time shifted Gaussian used by Bayliss et al. in ([3]) i.e.

$$(4.3) \qquad S(t) = \begin{cases} e^{-\alpha^2(t-t_s)^2} + e^{-\alpha^2 t_s^2}(t - t_s)^2\alpha^2 & 0 \leq t \leq 2t_s \\ 0 & 2t_s < t \end{cases}$$

where t_s is the time shift. The peak frequency f_{peak} of the Gaussian S(t) occurs at

$$(4.4) \qquad f_{peak} = \alpha/\pi \text{ Hz}$$

The maximum frequency of the wavelets, used to determine grid and time step requirements for a given model, was arbitrarily chosen to be three times the peak frequency f_{peak}.

Examples.

Lamb Problem. We consider now the two dimensional propagation of elastic waves in a uniform and isotropic half space excited by a point source located at $x = 0$. on the upper boundary free surface (at $y = -1$). The other 3 (artificial) computational boundaries are placed at $x = \pm 1$ and $y = +1$. This is a difficult numerical problem because of the presence of Rayleigh surface waves which propagate parallel to the grid and decay exponentially away from the free surface. Thus a stable boundary treatment and high accuracy are required near the free boundary.

In the present work the material parameters were taken to be $\rho = \lambda = \mu = 1$, so that the compression and shear wave speeds are $c_p = \sqrt{3}$ and $c_s = 1$, respectively. The source parameters were $\alpha = 8.31$ and $t_s = .25$, corresponding to a peak frequency of 2.6 Hz. The time history of the surface source function $S^{(1)}(t)$ is displayed in figure 1.

In figures 2a - 2c we compare computed solutions for the horizontal velocity u at a receiver located on the surface at $x = .5$. We have arbitrarily chosen to resolve waves up to a $3f_{peak}$ and consider results of the 2-4 finite difference scheme on a 400×400 grid to be the *exact* solution. This corresponds to a resolution of approximately 77 points per wavelength for the waves with frequency f_{peak} and 25 points per wavelength for waves with frequency $3f_{peak}$.

In figures 2a and 2b we compare this *exact* solution (solid line) with results of the 2-4 finite difference scheme (dashed lines) with resolutions of 50×50 and 100×100. In figure 2c

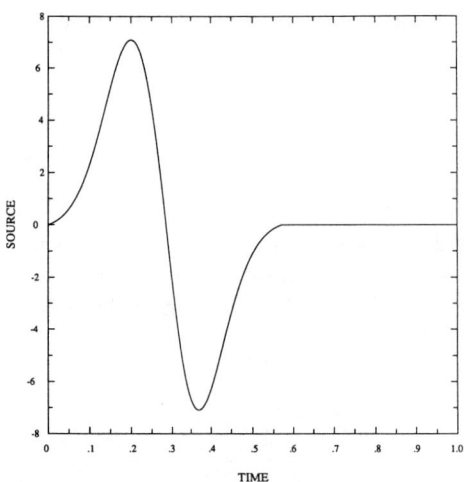

Figure 1. Time history of source function $S^{(1)}(t)$ for surface source.

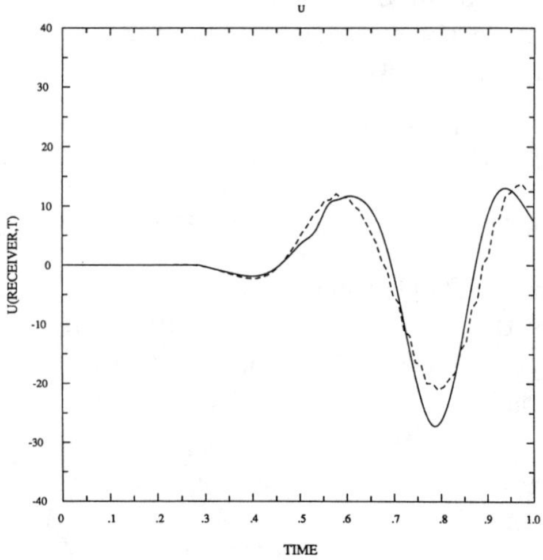

Figure 2a. Comparison of exact solution of Lamb problem (solid lines) to solution 2-4 finite difference solution using a grid of 50x50 (dashed lines). Horizontal velocity u plotted at a receiver location $x=.5$.

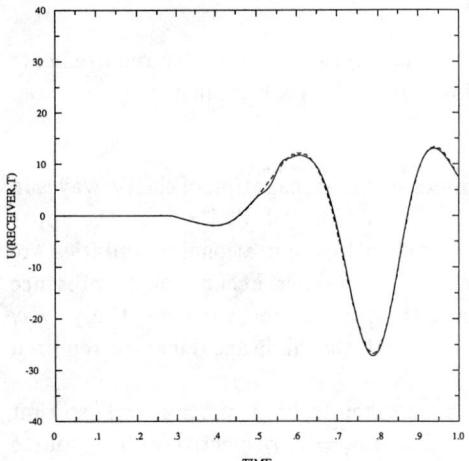

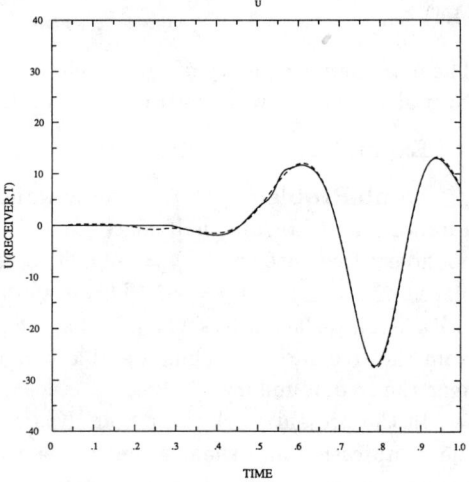

Figure 2b. Comparison of exact solution of Lamb problem (solid lines) to solution with 2-4 finite difference solution using a grid of 100x100 (dashed lines). Horizontal velocity u plotted at a receiver location $x=.5$.

Figure 2c. Comparison of exact solution of Lamb problem (solid lines) to solution with Chebyshev pseudospectral method on a grid with 55 x 50 collocation points. Horizontal velocity u plotted at a receiver location $x=.5$.

we compare the *exact* solution (solid line) with results from the Chebyshev method (dashed lines) with 55×50 collocation points. The spectral results are comparable to those of the 4th order finite difference scheme with twice the number of grid points in each coordinate direction.

In figure 3. we plot a snapshot of the pressure field $p(x, y) = .5(\tau_{11} + \tau_{22})$ at time $t = 1$. Here we see that the compressional body wave and the Rayleigh surface wave are well resolved.

Scattering Off a Plane Interface. We next consider the scattering of elastic waves off a plane interface. In this example we use a buried source whose time history is the second derivative of (4.3). The source is located at $x = 0.$, $y = -.5$. The source parameters are $\alpha = 11.3$, and $t_s = .285$, corresponding to a peak frequency of 3.6 Hz. The time history of the buried source function $S^{(2)}(t)$ is displayed in figure 4. This source was applied as a forcing term to the vertical momentum equation in (2.3) (ie. $f_1 = 0$, $f_2 = \delta(x)\delta(y + .5)S^{(2)}(t)$).

The material parameters were taken to be $\rho = \lambda = \mu = 1.$ above the interface and $\rho = \lambda = 1$, $\mu = 4$ below the interface. The wave speeds are therefore $c_p = \sqrt{3}$ and $c_s = 1$ above the interface and $c_p = 3.$ and $c_s = 2.$ below the interface. The interface is located at the center of the computational domain, at y=0.

As with the Lamb problem, we place a receiver along the free surface at $x = .5$ and compare solutions computed with the 2-4 finite difference scheme with those computed with the Chebyshev method. In figure 5a we plot the horizontal velocity u for solutions computed with the 2-4 finite difference scheme with a resolution of 200×200 (solid line) and 50×50 (dashed lines). After the arrival of the primary wave we can see amplitude and phase errors in the reflected and surface waves. In figure 5b we compare the horizontal velocity u for solutions computed with the 2-4 finite difference scheme with a resolution of 200×200 (solid line) and 100×100 (dashed lines). Here the results are almost identical, and we therefore consider the 200×200 computation to be the *exact* solution. Finally, in figure 5c we compare the *exact* horizontal velocity (solid lines) with the horizontal velocity computed by the Chebyshev method with 43×43 collocation points (dashed lines). Here we can see that the two solutions agree almost exactly until about $t = 1.2$. This time includes the arrival of both primary and reflected body waves. The arrival of the primary wave at the free surface sets off a surface Rayleigh wave. When the surface Rayleigh wave reaches the lateral boundary a global error is introduced since the boundary conditions do not absorb Rayleigh waves. Hence the high gradient that results from the steepening of the Rayleigh waves sets up a global Gibbs oscillation. This error shows up at every point on the surface immediately due to the incorrect boundary conditions. This error is also present in the finite difference scheme, however the error is not a global one as it propagates with a finite speed. The solutions are nevertheless inaccurate after this time. Thus up to the time when the surface Rayleigh waves reach a lateral boundary, the spectral method computes the propagation of internal body waves with an accuracy comparable to that of the fourth order finite difference scheme with more than twice the number of grid points in each coordinate direction.

5. Conclusions.

We have introduced a Chebyshev based pseudospectral method for elastic wave propagation suitable for limited area modeling. This method does not require periodic boundary conditions and hence realistic boundary conditions may be implemented. The free surface boundary condition appears to be quite stable and we have not observed any stability problems, even for calculations with large poisson ratios.

We have compared results of the Chebyshev pseudospectral method with those of a

SURFACE PLOT for P(X,Y)

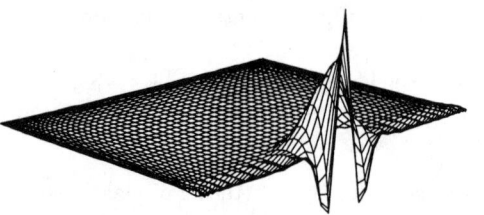

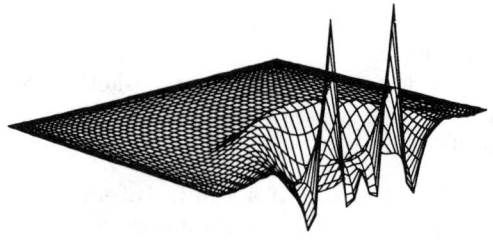

Figure 3a. Snapshot of pressure field for Lamb problem (as computed by Chebyshev spectral method) at time $t=.25$.

Figure 3b. Snapshot of pressure field for Lamb problem (as computed by Chebyshev spectral method) at time $t=.5$.

SURFACE PLOT for P(X,Y)

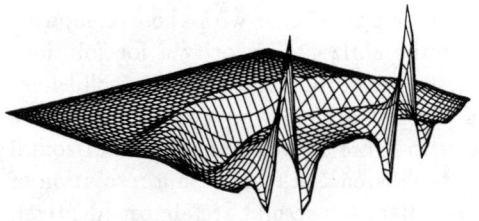

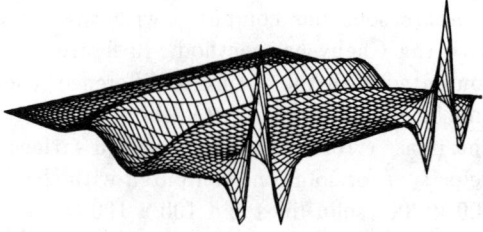

Figure 3c. Snapshot of pressure field for Lamb problem (as computed by Chebyshev spectral method) at time $t=.75$.

Figure 3d. Snapshot of pressure field for Lamb problem (as computed by Chebyshev spectral method) at time $t=1$.

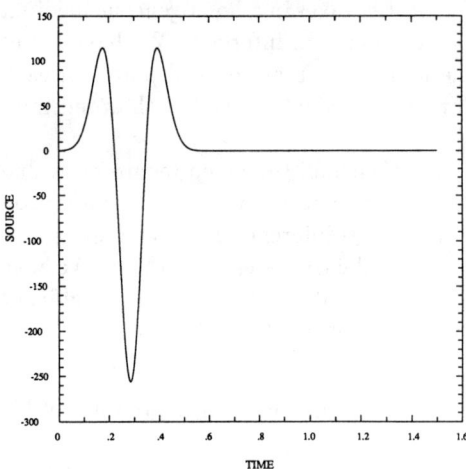

Figure 4. Time history of source function $S^{(2)}(t)$ for buried source.

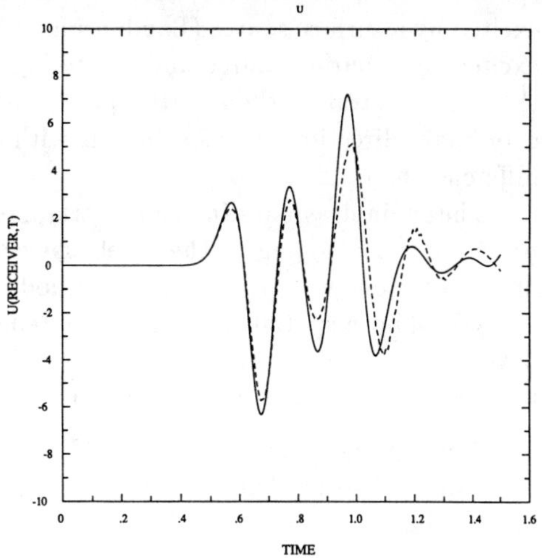

Figure 5a. Comparison of *exact* solution of Scattering problem (solid lines) to solution with 2-4 finite difference solution using a grid of 50x50 (dashed lines). Horizontal velocity u plotted at a receiver location $x=.5$.

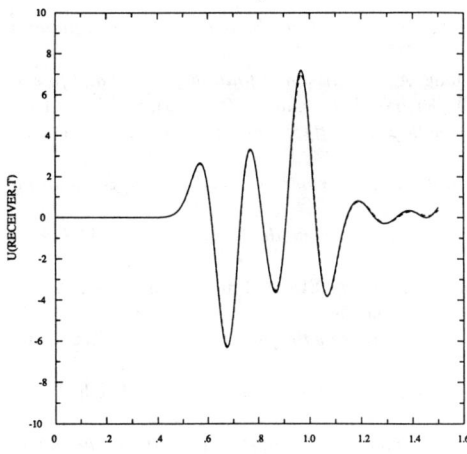

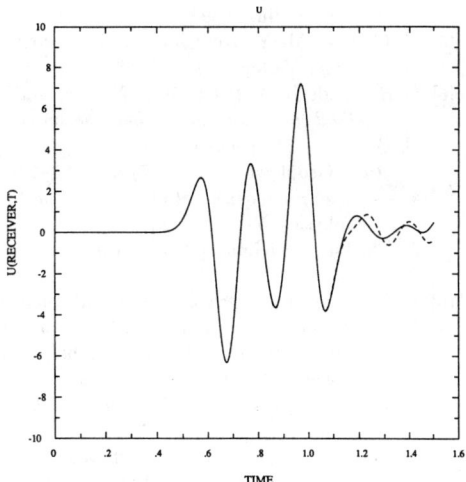

Figure 5b. Comparison of *exact* solution of Scattering problem (solid lines) to solution with 2-4 finite difference solution using a grid of 100x100 (dashed lines). Horizontal velocity u plotted at a receiver location $x=.5$.

Figure 5c. Comparison of *exact* solution of Scattering problem (solid lines) to solution with Chebyshev pseudospectral method on a grid with 43 x 43 collocation points. Horizontal velocity u plotted at a receiver location $x=.5$.

spatially fourth order finite difference method for elastic waves in a homogeneous medium excited by a surface source (Lamb problem) and for elastic waves in a homogeneous medium excited by a buried source and scattering off a plane horizontal interface. We have found that the spectral method, with approximately one half the number of grid points in each coordinate direction, yields solutions with an accuracy equivalent to the fourth order finite difference method .

The main drawback in the spectral method lies in the small time step requirement due to the ill conditioning of the Chebyshev differentiation matrices. A detailed comparison of the *efficiency* of the spectral method versus the finite difference method requires an analysis of both spatial resolution requirements as well as the time step restriction. We have developed a multidomain spectral method for the elastic equations which allows significant improvements in the time step restriction. This will be reported elsewhere [2].

Acknowledgments.

I would like to thank Alvin Bayliss for many useful discussions and for the use of his finite difference code.

REFERENCES

[1] J. M. Augenbaum, *Multidomain adaptive pseudospectral method for acoustic wave propagation in discontinuous media*, in Computational Acoustics-Seismo-Ocean Acoustics and Modeling, D. Lee, A. Cakmak, and R. Vichnevetsky, eds., Elsevier Science Publishers (North Holland), 1990, pp. 19–40.

[2] ———, *Multidomain pseudospectral method for elastic wave calculations in layered media.* (in preparation), 1991.

[3] A. Bayliss, K. E. Jordan, B. J. LeMesurier, and E. Turkel, *A fourth order accurate finite difference scheme for the computation of elastic waves*, Bull. Seismol. Soc. Amer., 76 (1986), pp. 1115–1132.

[4] J. P. Boyd, *Chebyshev and Fourier Spectral Methods*, vol. 49 of Lecture Notes in Engineering, Springer-Verlag, Berlin, 1989.

[5] C. Canuto, M. Y. Hussaini, A. Quarteroni, and T. A. Zang, *Spectral Methods in Fluid Dynamics*, Springer-Verlag, Berlin, 1988.

[6] C. R. Daudt, L. W. Braile, R. L. Nowack, and C. S. Chiang, *A comparison of finite-difference and fourier method calclations of synthetic seismograms*, Bull. Seismol. Soc. Amer., 79 (1989), pp. 1210–1230.

[7] B. Fornberg, *The pseudospectral method: comparisons with finite differences for the elastic wave equation*, Geophysics, 52 (1987), pp. 483–501.

[8] ———, *The pseudospectral method: accurate representation of interfaces in elastic wave calculations*, Geophysics, 53 (1988), pp. 625–637.

[9] J. Gazdag, *Modeling of the acoustic wave equation with transform methods*, Geophysics, 46 (1981), pp. 854–859.

[10] D. Gottlieb and S. Orszag, *Numerical Analysis of Spectral Methods*, SIAM, Philadelphia, PA., 1977.

[11] D. Kosloff and E. Baysal, *Forward modeling by a fourier method*, Geophysics, 47 (1982), pp. 1402–1412.

[12] D. Kosloff, M. Reshef, and D. Loewenthal, *Elastic wave calculations by the fourier method*, Bull. Seismol. Soc. Amer., 74 (1984), pp. 875–891.

[13] G. Raggio, *A pseudospectral chebyshev scheme for forward acoustic modeling*, Z. Angew. Math. Mech., 66 (1986), pp. 545–553.

[14] M. Reshef, D. Kosloff, D. Edwards, and C. Hsiung, *Three dimensional acoustic modeling by the fourier method*, Geophysics, 53 (1988), pp. 1175–1183.

[15] ———, *Three dimensional elastic modeling by the fourier method*, Geophysics, 53 (1988), pp. 1184–1193.

[16] G. Shubin and J. Bell, *A modified equation approach to constructing fourth order methods for acoustic wave propagation*, SIAM J. Sci. Stat Comput., 8 (1987), pp. 135–151.

[17] A. Solomonoff and E. Turkel, *Global properties of pseudospectral methods*, J. Comp. Phys., 81 (1989), pp. 239–276.

[18] R. Voigt, D. Gottlieb, and M. Y. Hussaini, eds., *Spectral Methods for Partial Differential Equations*, Philadelphia, PA., 1984, SIAM.

CHAPTER 2

Local Grid Refinement for Reservoir Simulation*

B. A. Boyett**
M. S. El-Mandouh**
R. E. Ewing†

Abstract. Local grid refinement in space and time in a three-dimensional, multiphase reservoir simulator is presented. A global rectangular coarse grid is refined locally with radial or rectangular patches around wells or other areas with important local flow properties.

For reservoir simulation, the need to use variable-sized time steps also arises. The time step size for the coarse grid is unsuitable for very fine grids because it can cause the model to produce unreasonable results or even fail to converge. To minimize this problem, local refinement in time is applied on the local fine-grid patches.

Iterative solvers for implementing local refinement into existing simulators will be discussed. Numerical results are presented from multiwell problems with time-varying rates.

1. Introduction

Reservoir simulation is used to understand and predict hydrocarbon recovery under a variety of production strategies. Most conventional simulators are based on dividing a reservoir into a quasi-uniform rectangular grid system and then solving the mass conservation equations that describe fluid flow between the grid blocks. Existing simulators of this form are suitable for modeling large-scale flow processes, but are frequently inadequate for describing smaller-scale phenomena such as flow into and around wells. Local grid refinement may be essential to resolve these fine local processes, and many schemes have been proposed for using local refinement to increase the accuracy of a simulation in areas in which more detail is needed to understand the reservoir behavior (eg. [10,16,17]). Unfortunately, these more accurate schemes can destroy the efficiency of existing solution algorithms in the simulations [16,17]. Since a great deal of time and money has been spent

*This research was supported in part by the Office of Naval Research Contract No. N00014-88-K-0370 and by the Institute for Scientific Computation through NSF Grant No. RII-8610680.
**Mobil Research and Development Corporation, 13777 Midway Road, Dallas, Texas 75234.
†Institute for Scientific Computation, University of Wyoming, P.O. Box 3036, Laramie, Wyoming 82071.

on developing and optimizing the existing conventional simulators, our primary focus must be on those methods in which the local refinement can be incorporated into these existing simulators while maintaining the efficiency of the original code.

To be practical and economical from an implementation standpoint, a local refinement strategy must meet certain criteria. The difference approximations used to represent fluid flow in the fine-grid zones and between the fine and coarse grid must be accurate and compatible with the cell-centered finite difference approximations used in most conventional simulators. Mass must be conserved across refinement interfaces. Also, since many of the processes we would like to study are dynamic, we need to be able to move and change the local fine grids with time. This requires relatively simple data structures to connect the fine and coarse grids, structures that do not generate a large overhead burden. Preliminary techniques for using local grid refinement techniques to follow moving fronts in multiphase flow problems appear in [4,5].

In addition, we must be able to refine in time as well as space. The time-step size in a simulation is limited by the size of the smallest grid block, and thus, the locally refined grid system can require a much smaller time step than the large coarse grid. Since using the small local time step for the entire reservoir can be prohibitively expensive, we must have some method in which we apply the small time steps only on the refined grid and a larger time step on the coarse grid, as well as an efficient solution algorithm for solving problems on the entire composite grid.

Another problem that has plagued large-scale reservoir simulators is the difficulty in treating local transients around wells in fully implicit codes. When one well is rapidly opened or shut to flow, the local fluid properties around the well change sufficiently quickly that the global Newton-Raphson method used to linearize the flow equations does not converge. The present industrial solution is to cut the time step over the whole reservoir to obtain convergence, even though the difficulty is highly localized. This is extremely wasteful, and computationally intensive. The domain decomposition techniques discussed here are very useful for developing accurate and efficient local time-stepping algorithms [11] or for getting better local initial guesses for the Newton-Raphson iteration.

In this paper, we describe a patch refinement strategy in which we combine either cylindrical grids around wells or fine rectangular grids around clustered wells with standard large-scale rectangular grids.

We focus on a domain-decomposition based preconditioner used to solve the linear problems resulting from the composite-grid system. Local time stepping is incorporated by using small time steps on the fine local grids to provide an initial guess for Newton's method for each coarse time step. We present results from a three-dimensional, three-phase black-oil simulator which incorporates this refinement strategy.

Solution procedure. The solution process on the composite grid system is similar to that employed in conventional simulators with standard rectangular grid systems. We first linearize the flow equations using Newton's method and then solve the resulting linear problem using a preconditioned generalized conjugate residual (PGCR) iteration [7]. For composite-grid applications, the process is combined with a domain-decomposition based preconditioner developed by Bramble, Ewing, Pasciak, and Schatz (BEPS) [2]. Local time stepping on the fine local patches can be used if necessary.

In the local time-step option, each global time step consists of two parts: (a) local solutions on the fine local grids at intermediate time within each coarse-grid time step, and (b) a global solution on the entire composite grid at the coarse time-step level. First, for each fine-grid region, the global coarse-grid time step is divided into smaller time steps which are suitable for that region. The local fine-grid problems are solved for each of the smaller time steps using the old time-step solution on the coarse grid as a fixed boundary condition. This results in a new solution on each local fine-grid system at the new coarse time. Newton's method is then applied at the new coarse time step on the composite grid using the new solution on the local fine-grid patches and the old solution on the coarse grid for an initial guess. The resulting linear problem is then solved iteratively using PGCR, preconditioned with the BEPS procedure described below.

The primary attraction of the BEPS preconditioner is the ease of implementing it into a conventional simulator. BEPS requires matrix solutions only for coarse grid problems and local subgrid problems. In an existing simulator, the linear solution algorithm for solving coarse-grid matrix problems is already in place and is usually effectively vectorized. Therefore, to implement the BEPS procedure, all that is required is to add solutions for the local subgrid problems. In this sense then, no major modifications are needed in the existing simulator.

2. Multiphase Flow Equations

In order to understand the complexities of incorporating local grid refinement capabilities into existing multiphase simulators, we first review the basic properties of these simulators. Since the conservation equations for multiphase fluid flow in porous media are well known [1,4,15], these will not be rederived here. Instead, for sake of completeness, the three-phase, black-oil, flow equations that are used in our computational examples will be listed below. The equations for conservation of the oil and water phases are given by $(i = o, w)$

$$\nabla \cdot [\lambda_i (\nabla p_i - \rho_i g \nabla z)] + q_i = \frac{\partial (\phi B_i S_i)}{\partial t}. \tag{2.1}$$

The gas equation is

$$\nabla \cdot [\lambda_g (\nabla p_g - \rho_g g \nabla z)] + \nabla \cdot [R_s \lambda_o (\nabla p_o - \rho_o g \nabla z)] + q_q + R_s q_o$$

$$= \frac{\partial (\phi B_q S_q)}{\partial t} + \frac{\partial (R_s \phi B_o S_o)}{\partial t}. \tag{2.2}$$

The saturations are related by

$$S_o + S_w + S_g = 1, \tag{2.3}$$

and the pressures are related to the capillary pressures by

$$p_{cow} = p_o - p_w, \tag{2.4}$$

$$p_{cgo} = p_g - p_o. \tag{2.5}$$

These capillary pressure relationships are strongly nonlinear functions of the phase saturations. Darcy's law is given by

$$v_i = -\lambda_i \nabla \Phi_i \equiv -k \frac{k_{ri}}{B_i \mu_i} \nabla \Phi_i, \tag{2.6}$$

with $i = o, w, g$, where k is the permeability tensor, v_i is the ith phase velocity, and the flow potential gradient is defined, for $i = o, w, g$, as

$$\nabla \Phi_i = \nabla p_i - \rho_i g \nabla z. \tag{2.7}$$

Again, the relative permeabilities k_{ri}, which describe the competition of the fluid phases for flow, are nonlinear functions of the phase saturations.

The flow rate across any boundary surface A may be expressed in terms of Φ_i as

$$Q_i \bigg|_A = \int \int_A k \frac{k_{ri}}{B_i \mu_i} (\nabla \Phi_i \cdot n) \, dA \tag{2.8}$$

for $i = o, w, g$. We obtain accurate finite difference formulae by conserving mass across grid interfaces through matching Q from (2.8) on both sides of the face. See [12,13] for cell-centered finite difference formulations on the composite grids that are accurate and that conserve mass. Once a composite-grid operator is defined for accuracy, its efficient solution must be addressed.

3. Efficient Solution Methods

We first consider a regular, quasi-uniform grid on Ω, the computational domain. We consider the matrix $\mathbf{A}^c$, generated by a finite element or finite difference approximation of the flow equations using a coarse, quasi-uniform mesh. Let $\Omega_1 \subset \Omega$ be a region that contains some important local phenomenon that may require better resolution. Denote $\Omega \setminus \Omega_1$ by Ω_2. Then Ω_1 and Ω_2 produce a natural decomposition of Ω. Let the solution vector $\mathbf{x}^c$ of the original coarse-grid problem be decomposed in the form $\mathbf{x}^c = (x_1^c, x_2^c)^T$ where x_1^c and x_2^c are the parts of the solution in Ω_1 and Ω_2, respectively. The corresponding decomposition of the matrix $\mathbf{A}^c$ can be described in

$$\mathbf{A}^c \mathbf{x}^c = \begin{pmatrix} A_{11}^c & A_{12}^c \\ A_{21}^c & A_{22}^c \end{pmatrix} \begin{pmatrix} x_1^c \\ x_2^c \end{pmatrix}. \tag{3.1}$$

We assume that a code exists or can easily be written to solve (3.1) for a quasi-uniform grid; the code can take advantage of the banded structure of the matrix $\bar{\mathbf{A}}^c$, which is equivalent to $\mathbf{A}^c$, but which utilizes a standard lexicographical ordering of the unknowns. Without loss of generality, we can assume that $\mathbf{x}^c$ represents a difference from a fixed value of the unknown, as in the difference of the vector over a time step in a transient problem and the right-hand side is the corresponding difference in right-hand sides of the original problem. This practice of solving for vector changes can significantly reduce round-off errors in large-scale computations. Of course, due to the nonlinearities involved in our flow equations, we assume that the linear problems discussed here are the results of some Newton-Raphson outer iteration.

Next, we assume that due to the presence of some localized phenomenon, grid refinement is desired in Ω_1. Let x_r be the new approximation on the refined grid in Ω_1 and A_{rr} be the local matrix in Ω_1. Let A_{r2} and A_{2r} be the new connection matrices between the new unknowns in Ω_1 and those in the original coarse grid in Ω_2. Then, in order to maintain a simple data structure, we add the new unknowns x_r to the beginning of the vector $\mathbf{x}^c$ to obtain a new vector $\tilde{\mathbf{x}}^c = (x_r, x_1^c, x_2^c)^T$. We can write the composite matrix problem in the form

$$\tilde{\mathbf{A}} \tilde{\mathbf{x}} \equiv \begin{pmatrix} A_{rr} & 0 & A_{r2} \\ 0 & I & 0 \\ A_{2r} & 0 & A_{22}^c \end{pmatrix} \begin{pmatrix} x_r \\ x_1^c \\ x_2^c \end{pmatrix} = \begin{pmatrix} b_1 \\ 0 \\ b_2 \end{pmatrix} \equiv \tilde{\mathbf{b}}, \tag{3.2}$$

or, by eliminating the x_1^c unknowns completely and redefining the r subscripts and corresponding new variables, (3.2) can be written as

$$\mathbf{A} \mathbf{x} \equiv \begin{pmatrix} A_{11} & A_{12} \\ A_{21} & A_{22} \end{pmatrix} \begin{pmatrix} x_1 \\ x_2 \end{pmatrix} = \begin{pmatrix} b_1 \\ b_2 \end{pmatrix} \equiv \mathbf{b}. \tag{3.3}$$

Here $A_{rr} = A_{11}$, $A_{r2} = A_{12}$, $A_{2r} = A_{21}$, $A_{22}^c = A_{22}$, $x_1 = x_r$, and $x_2 = x_2^c$, are all defined using the composite grid.

Thus, we want to solve the composite grid problem $\mathbf{Ax} = \mathbf{b}$, defined by (3.2) and (3.3). The composite grid unknowns can not be reordered to obtain a banded matrix that is equivalent to $\mathbf{A}$, as $\mathbf{A}^c$ could be rearranged to obtain $\bar{\mathbf{A}}^c$. Also, the matrix $A_{22} = A_{22}^c$ can not be rearranged to obtain a banded matrix for ease of vector solution.

A common way to iteratively solve (3.3) is to use a block Gauss-Seidel (BGS) iterative method, where A_{11} and A_{22} are inverted sequentially with previous iterates passed to the right-hand side with A_{12} and A_{21}. This breaks apart the important information that connects the unknowns in Ω_1 and Ω_2 in the inversion process and results in very slow convergence properties for many applications. In the domain decomposition terminology, this is equivalent to an ordinary Schwarz alternating procedure between Ω_1 and Ω_2 with no overlap.

We note that $\mathbf{A}$ can be factored into triangular block matrices of the form

$$
\begin{pmatrix} A_{11} & A_{12} \\ A_{21} & A_{22} \end{pmatrix} = \begin{pmatrix} A_{11} & 0 \\ A_{21} & A_{22} - A_{21}A_{11}^{-1}A_{12} \end{pmatrix} \begin{pmatrix} I & A_{11}^{-1}A_{12} \\ 0 & I \end{pmatrix}
$$

$$
= \begin{pmatrix} A_{11} & 0 \\ A_{21} & S \end{pmatrix} \begin{pmatrix} I & A_{11}^{-1}A_{12} \\ 0 & I \end{pmatrix},
$$

(3.4)

where $S = A_{22} - A_{21}A_{11}^{-1}A_{12}$ is the Schur complement of A_{11} and I denotes the block identity matrix of the size corresponding to the matrix decomposition in each position. Similarly, we see that $\mathbf{A}^{-1}$ can be written as

$$
\begin{pmatrix} A_{11} & A_{12} \\ A_{21} & A_{22} \end{pmatrix}^{-1} = \begin{pmatrix} I & A_{11}^{-1}A_{12} \\ 0 & I \end{pmatrix}^{-1} \begin{pmatrix} A_{11} & 0 \\ A_{21} & S \end{pmatrix}^{-1}
$$

$$
= \begin{pmatrix} I & -A_{11}^{-1}A_{12} \\ 0 & I \end{pmatrix} \begin{pmatrix} A_{11}^{-1} & 0 \\ -A_{11}^{-1}S^{-1}A_{21} & S^{-1} \end{pmatrix}.
$$

(3.5)

As was mentioned earlier, S is not generally an easy matrix to invert efficiently. However, since $A_{22} = A_{22}^c$, one can show that $S^c = A_{22}^c - A_{21}^c A_{11}^{-1} A_{12}$ is comparable to S with comparability constants independent of the mesh spacing Δx [12,13]. Although S^c is not an easy matrix to invert, if we solve the coarse-grid matrix problem generated by $\mathbf{A}^c$ by any method, the resulting solution x_2^c will be the same as the corresponding solution obtained by inverting S^c. Thus, we can obtain the action of $(S^c)^{-1}$ by a solution process on $\mathbf{A}^c$, which we have assumed is a known or easy process. Therefore, instead of finding $\mathbf{A}^{-1}$ from (3.5), we can replace $\mathbf{A}$ by a preconditioning matrix $\mathbf{B}$, which is obtained from $\mathbf{A}$ by replacing S in (3.4) by S^c. The action of $\mathbf{B}^{-1}$ can then be obtained from

$$
\mathbf{B}^{-1} = \begin{pmatrix} I & -A_{11}^{-1}A_{12} \\ 0 & I \end{pmatrix} \begin{pmatrix} A_{11}^{-1} & 0 \\ -A_{11}^{-1}S^{c-1}A_{21} & S^{c-1} \end{pmatrix}.
$$

(3.6)

This is the preconditioner defined by Bramble, Ewing, Pasciak, and Schatz [2], which has been called the BEPS preconditioner.

We can use (3.6) to define a stand-alone iterative process as in [6], or we can use (3.6) as a preconditioner in a preconditioned conjugate gradient PCG (or residual, PCR) iterative method. We will first describe the pure iterative procedure. This procedure will be uniformly well conditioned for finite element or point-distributed finite difference discretizations that satisfy a variational principle, but not for cell-centered finite differences [12,13]. We see from our computational experiments [8,9] that these procedures may require scaling via an acceleration parameter τ to maintain rapid convergence.

a. Stand-Alone Iterative Procedure. We can first solve the composite-grid problem (3.3) with a zero initial guess for x_2 related to zero Dirichlet conditions on the boundary of Ω_1 to obtain our first iterate x_1^0

$$x_1^0 = A_{11}^{-1} b_1. \tag{3.7}$$

This problem can be solved exactly, or approximately, by some iterative technique. If the fine-grid problems are solved exactly, we will perform the iteration in a restricted subspace, which greatly reduces the number of iterations required to obtain a specific error tolerance. Thus, given the starting point

$$\mathbf{x}^0 = \begin{pmatrix} A_{11}^{-1} b_1 \\ 0 \end{pmatrix}, \tag{3.8}$$

we can define our iterative method for the correction $\mathbf{c}^n$ to $\mathbf{x}^0$. Using the approximation $\mathbf{x}^0$, we want to obtain an approximation for $\mathbf{x}^n$ given by $\mathbf{x}^n = \mathbf{x}^0 + \mathbf{c}^n$; thus $\mathbf{c}^0 = 0$.

We now define an iterative procedure using (3.6) and an iteration parameter τ by:

$$\mathbf{c}^{n+1} = \tau \mathbf{B}^{-1} (\mathbf{b} - \mathbf{A}\mathbf{x}^n), \qquad n = 1, 2, \cdots, \tag{3.9}$$

where $\mathbf{B}^{-1}$ is defined in (3.6). Consider the residual vector defined from the composite matrix (3.3) by

$$
\begin{aligned}
\mathbf{r}^n &= \mathbf{b} - \mathbf{A}\mathbf{x}^n = \mathbf{b} - \mathbf{A} (\mathbf{x}^0 + \mathbf{c}^n) = \mathbf{b} - \mathbf{A}\mathbf{x}^0 - \mathbf{A}\mathbf{c}^n \\
&= \begin{pmatrix} b_1 \\ b_2 \end{pmatrix} - \begin{pmatrix} A_{11} & A_{12} \\ A_{21} & A_{22} \end{pmatrix} \begin{pmatrix} A_{11}^{-1} b_1 \\ 0 \end{pmatrix} - \begin{pmatrix} A_{11} & A_{12} \\ A_{21} & A_{22} \end{pmatrix} \begin{pmatrix} c_1^n \\ c_2^n \end{pmatrix} \\
&= \begin{pmatrix} 0 \\ b_2 - A_{21} A_{11}^{-1} b_1 \end{pmatrix} - \begin{pmatrix} A_{11} & A_{12} \\ A_{21} & A_{22} \end{pmatrix} \begin{pmatrix} c_1^n \\ c_2^n \end{pmatrix} \\
&\equiv \mathbf{d} - \mathbf{A}\mathbf{c}^n,
\end{aligned}
\tag{3.10}
$$

where $\mathbf{d} = (0, d_2)^T$ has a special form with $d_1 = 0$ if (3.7) is solved exactly. Thus, in this case, we can redefine the iterative method to find $\mathbf{x}^{n+1} = \mathbf{x}^0 + \mathbf{c}^{n+1}$ that satisfies

$$\mathbf{c}^{n+1} = \tau \mathbf{B}^{-1} (\mathbf{d} - \mathbf{A}\mathbf{c}^n), \qquad n = 1, 2, \cdots \tag{3.11}$$

Note that, from (3.6), we see that with $\mathbf{c}^0 = 0$,

$$
\begin{aligned}
\mathbf{B}^{-1}\mathbf{d} &= \begin{pmatrix} I & -A_{11}^{-1} A_{12} \\ 0 & I \end{pmatrix} \begin{pmatrix} A_{11}^{-1} & 0 \\ -A_{11}^{-1} S^{c-1} A_{21} & S^{c-1} \end{pmatrix} \begin{pmatrix} 0 \\ d_2 \end{pmatrix} \\
&= \begin{pmatrix} I & -A_{11}^{-1} A_{12} \\ 0 & I \end{pmatrix} \begin{pmatrix} 0 \\ S^{c-1} d_2 \end{pmatrix} \\
&= \begin{pmatrix} -A_{11}^{-1} A_{12} S^{c-1} d_2 \\ S^{c-1} d_2 \end{pmatrix}.
\end{aligned}
\tag{3.12}
$$

Therefore, we see that the action of $\mathbf{B}^{-1}$ on a vector of the form $(0, y_2)^T$ will only require the solution of one coarse-grid problem (to obtain the action of $S^{c^{-1}}$) and one fine-grid problem (to obtain the action of A_{11}^{-1}). We note explicitly that (formally) we have

$$c_2^1 = \tau S^{c^{-1}} d_2. \tag{3.13}$$

Then, we obtain

$$c_1^1 = -\tau A_{11}^{-1} A_{12} c_2^1. \tag{3.14}$$

Following the same arguments as in (3.12), we see that if we solve

$$\mathbf{A}^c \mathbf{c}^1 = \begin{pmatrix} 0 \\ d_2 \end{pmatrix} = \mathbf{d}, \tag{3.15}$$

where $\mathbf{d}$ is given by (3.11), we will obtain the solution c_2^1 given by (3.13) without the acceleration parameter τ. Then, given c_2^1, we transfer that information to the boundary of Ω_1 by $A_{12} c_2^1$ and solve a fine-grid problem to obtain the c_1^1 given by (3.14).

Next, we note that given $\mathbf{c}^1$ from (3.12), if we multiply it by $\mathbf{A}$, we obtain

$$
\begin{aligned}
\mathbf{A} \mathbf{c}^1 &= \begin{pmatrix} A_{11} & A_{12} \\ A_{21} & A_{22} \end{pmatrix} \begin{pmatrix} -A_{11}^{-1} A_{12} c_2^1 \\ c_2^1 \end{pmatrix} \tau \\
&= \begin{pmatrix} 0 \\ \left(-A_{21} A_{11}^{-1} A_{12} + A_{22} \right) c_2^1 \end{pmatrix} \tau = \begin{pmatrix} 0 \\ S S^{c^{-1}} d_2 \end{pmatrix} \tau.
\end{aligned}
\tag{3.16}
$$

Thus, each residual of the reduced iterative system will remain in the subspace with first component zero if (3.14) is solved exactly. We note that

$$\mathbf{d} - \mathbf{A} \mathbf{c}^1 = \begin{pmatrix} 0 \\ d_2 - S S^{c^{-1}} d_2 \tau \end{pmatrix} \tag{3.17}$$

and each succeeding residual will be in this space. Clearly, this residual measures the error incurred by using S^c instead of S in our procedure; it would be zero if S were used. Since S and S^c are comparable, independent of Δx, the convergence rate does not degrade with Δx.

We noted that if the local fine-grid problems are not solved exactly, then $\mathbf{d} - \mathbf{A} \mathbf{c}^n$ does not have zero as its first component. In this case, the action of $\mathbf{B}^{-1}$ given in (3.12) will require an extra solution of a fine-grid problem per iteration. Preliminary computations [9] indicate that, for cell-centered finite difference discretizations, if the subregion is solved iteratively with its own preconditioner, the full algorithm with two subregion solves will converge more slowly than the method with direct solves unless the iterative method on the subdomain reduces the error sufficiently. Iterative solution of the coarse-grid problem causes no difficulty with either version of the algorithm. This is an important consideration for the reservoir simulation applications when iterative solution of the coarse-grid problem is essential due to its size, since direct solution of the coarse-grid problem is usually not possible.

This iterative procedure is uniformly well-conditioned for finite element procedures such as those used in this paper or point-centered finite difference methods, but not for cell-centered finite differences [12,13]. For discretizations arising from cell-centered finite difference methods, scaling of the iteration via the parameter τ in (3.9) may be necessary.

Alternatively, one can use an iterative method like a preconditioned conjugate residual method with BEPS as a preconditioner without a need for scaling [12,13].

b. Preconditioned Gradient-type Iterative Methods. We briefly present the preconditioned conjugate gradient iterative method presented by Bramble, Ewing, Pasciak, and Schatz (BEPS [2]). Given the similarity with the iterative method described above, we explain the BEPS method via algebraic terminology presented earlier. Again, we solve the problem $\mathbf{Ax} = \mathbf{b}$ given by (3.3) using the preconditioner described in (3.6) and an iterative method like preconditioned conjugate gradients or preconditioned conjugate residuals. Given the partitioning between Ω_1 and Ω_2 as before, we define the general BEPS-preconditioner as follows:

1. solve a fine-grid problem on Ω_1:

$$A_{11}x^0 = b_1; \tag{3.18}$$

2. compute the defect:

$$\mathbf{d} = \mathbf{b} - \mathbf{A}\left(\begin{pmatrix} \mathbf{x}^0 \\ 0 \end{pmatrix} + \mathbf{c}\right) = \begin{pmatrix} 0 \\ b_2 - A_{21}A_{11}^{-1}b_1 \end{pmatrix}; \tag{3.19}$$

3. solve a coarse-grid problem using the right-hand side of (3.19) on the coarse grid for $\mathbf{c} = (c_1, c_2)^T$:

$$\mathbf{A}^c\mathbf{c} = \tilde{\mathbf{d}} \tag{3.20}$$

(here $\tilde{\mathbf{d}}$ corresponds to $\mathbf{d}$ restricted to the coarse grid);

4. find the fine-grid correction, using c_2 from (3.20):

$$A_{11}\tilde{c}_1 = -A_{12}c_2; \tag{3.21}$$

then

$$\mathbf{x} = \mathbf{B}^{-1}\mathbf{b} = \begin{pmatrix} x_0 + \tilde{c}_1 \\ c_2 \end{pmatrix}. \tag{3.22}$$

This algorithm corresponds to one application of the preconditioner $\mathbf{B}^{-1}$. The residual is formed using the composite matrix $\mathbf{A}$ so that (3.3) is solved. The method is then accelerated by some gradient-type procedure. The general application of $\mathbf{B}^{-1}$ then requires two fine-grid solves and one coarse-grid solve. As before, if A_{11} is inverted exactly in each case, x^0 is computed only once and one stays in the subspace with first component zero while computing $\tilde{\mathbf{c}} = (\tilde{c}_1, c_2)^T$ with only one inversion of A_{11} per iteration.

4. Computational Examples

The two examples illustrate two different applications for local grid refinement. The first shows how radial refinement with local time stepping can be used to increase resolution of flow around isolated wells. In the second example, rectangular refinement around a cluster of wells is used to reduce computer time and cost while maintaining the accuracy provided by a conventional simulator.

Example 1. The first example is a two-well gas-oil problem based on the SPE comparative solution for black oil systems [14]. The coarse grid is 6 × 6 × 3 with each block having areal dimensions of 500 ft. × 500 ft. An oil producer is located in one corner of the reservoir with production from all three layers, and a gas injector is placed in the top layer of the opposite corner of the reservoir. Radial refinement is used around both wells with ring boundaries located at 10, 100, and 200 ft. Because the refinement introduces such small grid blocks adjacent to the wells, a time-step size suitable for the coarse grid would be too large for the fine radial grids. For this reason, local time stepping is applied to obtain an initial guess for each global composite-grid time step.

Results from this problem are plotted in Figures 1 and 2. Figure 1 displays the gas-oil ratio at the producer as a function of time. As expected, the locally-refined solution shows earlier gas breakthrough than the coarse-grid solution because of the smaller cells immediately adjacent to the well. Bottomhole pressure at the producer is plotted in Figure 2 and also shows clear differences between the fine- and coarse-grid solutions.

Example 2. The second problem is a three-layer oil reservoir surrounded by an aquifer. Four producers are located in the reservoir. More resolution is required in the hydrocarbon zone around the wells than in the surrounding aquifer. In a conventional simulation, however, the fine gridding around the wells must be extended throughout the reservoir (Figure 3). The locally-refined grid (Figure 4) maintains the same degree of resolution around the wells, but with a smaller number of grid blocks. The conventional grid is a $7 \times 8 \times 3$ rectangular grid with a total of 168 grid blocks, while the locally-refined composite grid consists of a $4 \times 3 \times 3$ coarse grid with a $4 \times 6 \times 3$ rectangularly refined zone for a total of 108 grid blocks. Results from the two problems are shown in Figures 3 through 10 and agree quite well. Because of the smaller number of grid blocks, however, the locally-refined grid solution reduces computer time by 52% over the coarse-grid solution.

5. Conclusions

1. Local grid refinement can improve accuracy and save computer time and cost.
2. Patch refinement can be incorporated into large reservoir simulators without destroying the efficiency of the original code.
3. Local time stepping used on the fine local grids to provide an initial guess for Newton's method for each coarse time step avoids having to take prohibitively expensive small time steps over the entire reservoir.
4. The BEPS preconditioner requires matrix solutions only for the coarse grid problem and local subgrid problems . Therefore, to implement it in an existing simulator, all that is required is to add solutions for the local subgrid problems to the existing linear solution algorithm for solving the coarse grid systems.

Acknowledgments

This research was supported in part by the Office of Naval Research Contract No. N00014–88–K–0370 and by the Institute for Scientific Computation through NSF Grant No. RII–8610680. The authors wish to thank Mobil Research and Development for allowing us to publish this paper.

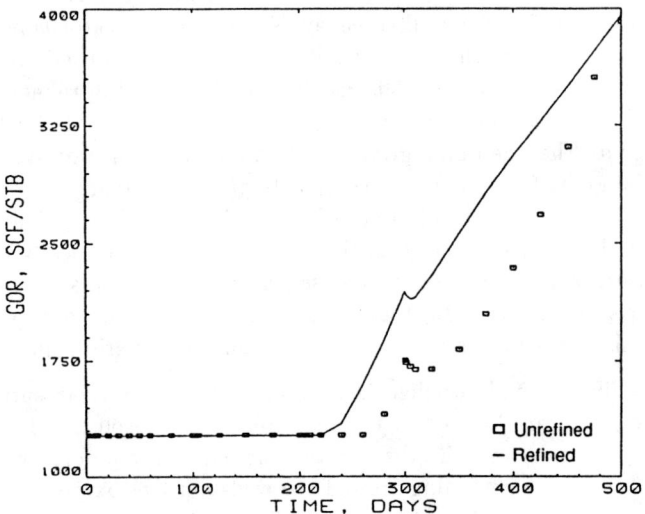

FIGURE 1. *Production gas-oil ratio for Example 1.*

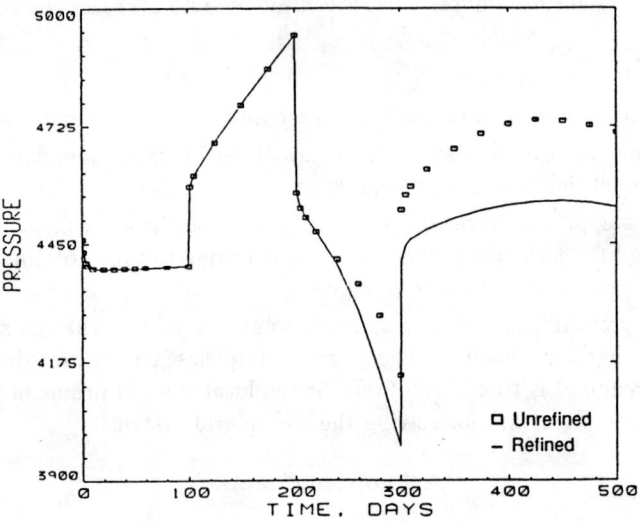

FIGURE 2. *Producing bottomhole pressure for Example 1.*

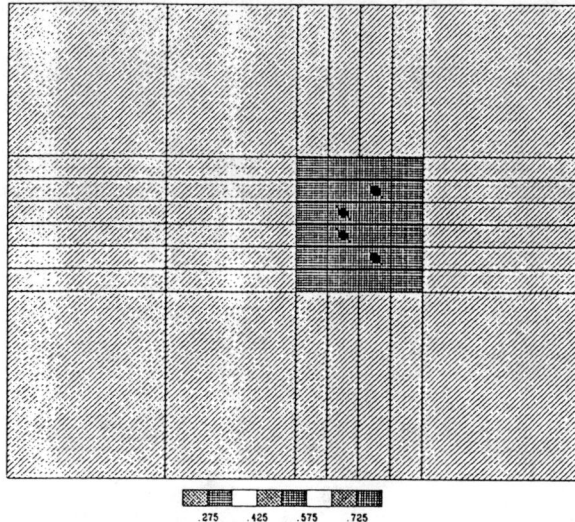

FIGURE 3. *Initial oil saturation on conventional grid for Example 2.*

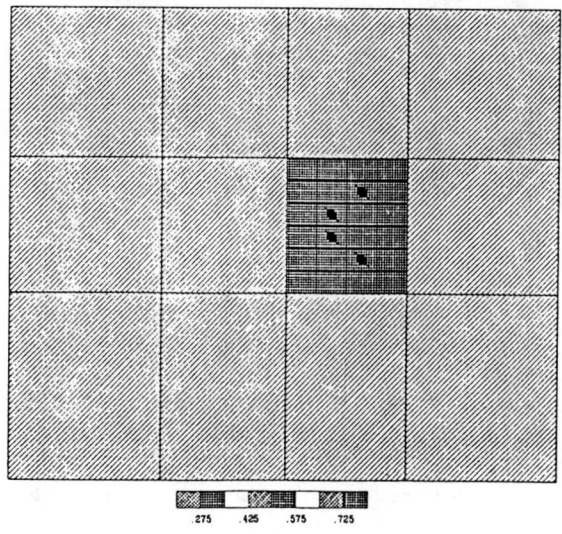

FIGURE 4. *Initial oil saturation on locally refined grid for Example 2.*

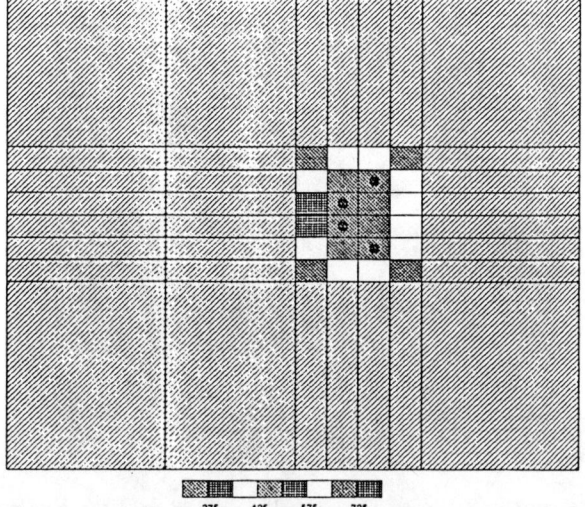

FIGURE 5. *Final oil saturation on conventional grid for Example 2.*

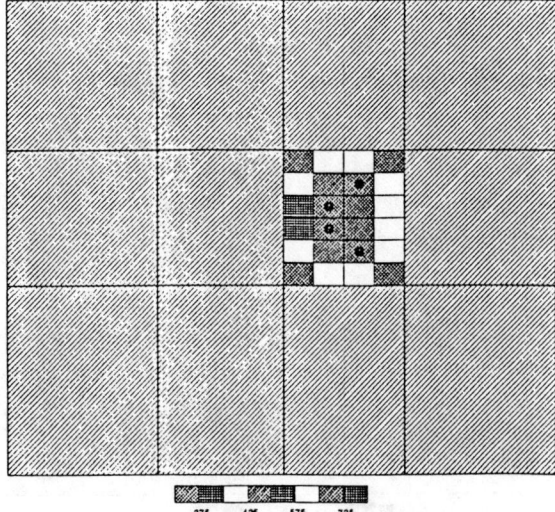

FIGURE 6. *Final oil saturation on locally refined grid for Example 2.*

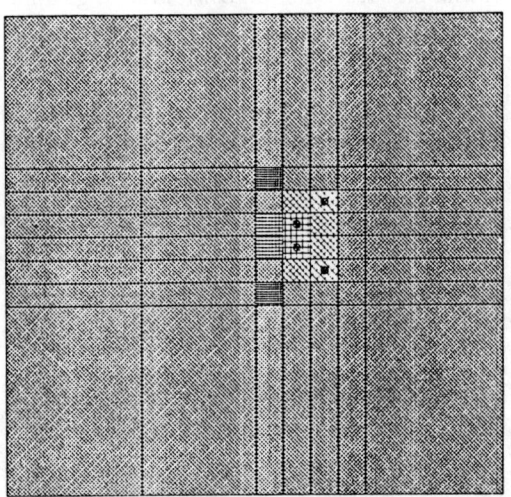

FIGURE 7. *Final pressure on conventional grid for Example 2.*

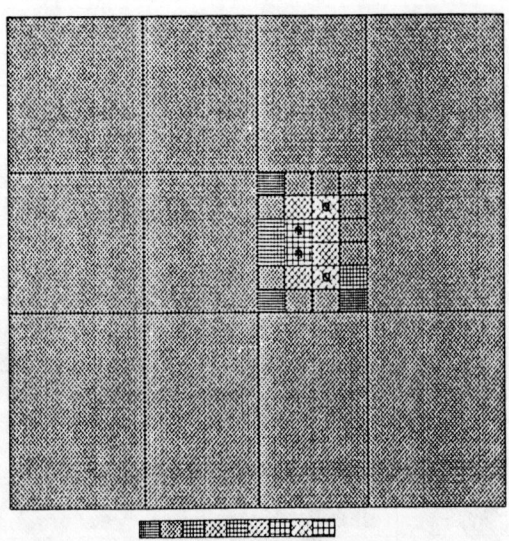

FIGURE 8. *Final pressure on locally refined grid for Example 2.*

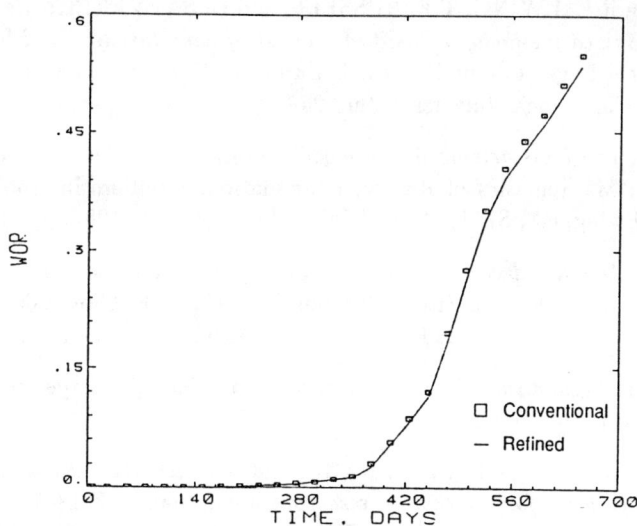

FIGURE 9. *Water oil ratio for Example 2.*

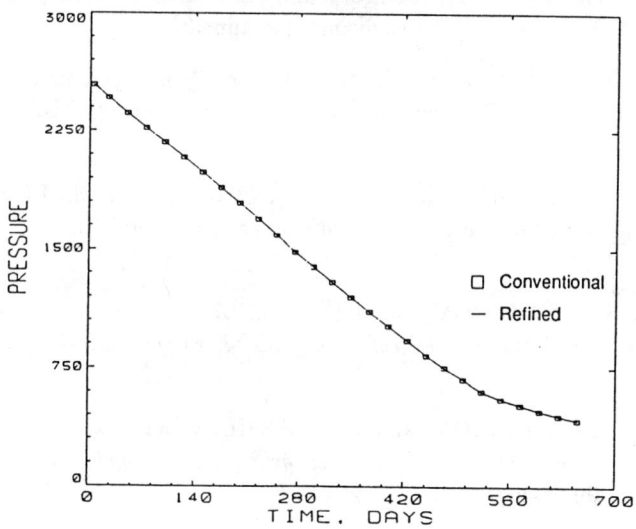

FIGURE 10. *Bottomhole pressure for Example 2.*

References

1. K. AZIZ and A. SETTARI, *Petroleum Reservoir Simulation*, Applied Science Publishers, London, 1979.

2. J.H. BRAMBLE, R.E. EWING, J.E. PASCIAK, and A.H. SHATZ, *A preconditioning technique for the efficient solution of problems with local grid refinement*, Computer Methods in Applied Mechanics and Engineering, 67 (1988), pp. 149–159.

3. M. ESPEDAL and R.E. EWING, *Characteristic Petrov-Galerkin subdomain methods for two-phase immiscible flow*, Computer Methods in Applied Mechanics and Engineering, 64 (1987), pp. 112–115.

4. M. ESPEDAL, R.E. EWING, T.F. RUSSELL, and O. SAEVAREID, *Reservoir simulation using mixed methods, a modified method of charcteristics and local grid refinement*, Proceedings of Joint IMA/SPE European Conference on Mathematics of Oil Recovery, Cambridge University, July 25–27, 1989, (to appear).

5. R.E. EWING, *Problems arising in the modeling of processes for hydrocarbon recovery*, in Vol. 1, Mathematics of Reservoir Simulation, Frontiers in Applied Mathematics, R.E. Ewing, ed., SIAM, Philadelphia, Pennsylvania, 1984, pp. 3–34.

6. R.E. EWING, *Domain decomposition techniques for efficient adaptive local grid refinement*, in Domain Decomposition Methods, T.F. Chan, R. Glowinski, T. Periaux, and O.B. Widlund, eds., SIAM, Philadelphia, Pennsylvania, 1989, pp. 192–206.

7. R.E. EWING, *Preconditioned conjugate gradient methods for large-scale fluid flow applications*, BIT, (to appear).

8. R.E. EWING, B.A. BOYETT, D.K. BABU, and R.F. HEINEMANN, *Efficient use of locally refined grids for multiphase reservoir simulation*, SPE 18413, Proceedings Tenth SPE Symposium on Reservoir Simulation, Houston, Texas, February 6–8, 1989, pp. 55–70.

9. R.E. EWING, P.G. JACOBS, R.R. PARASHKEVOV, and J. SHEN, *Applications of adaptive grid-refinement methods*, Proceedings of IIMAS Workshop on Numerical Methods, SIAM, Philadelphia, Pennsylvania, (to appear).

10. R.E. EWING and R.D. LAZAROV, *Adaptive local grid refinement*, paper SPE 17806, presented at the SPE Rocky Mountain Regional Meeting, Casper, Wyoming, May 11–13, 1988.

11. R.E. EWING, R.D. LAZAROV, J.E. PASCIAK, and P.S. VASSILEVSKI, *finite element methods for parabolic problems with timesteps variable in space*, (in preparation).

12. R.E. EWING, R.D. LAZAROV, and P.S. VASSILEVSKI, *Local refinement techniques for elliptic problems on cell-centered grids, I: Error analysis*, Math. Comp., (submitted).

13. R.E. EWING, R.D. LAZAROV, and P.S. VASSILEVSKI, *Local refinement techniques for elliptic problems on cell-centered grids, II: Two-grid iterative methods*, Math. Comp., (submitted).

14. A.S. ODEH, *Comparison of solutions to a three-dimensional black-oil reservoir simulation problem*, JPT, 33(13) (1981).

15. D.W. PEACEMAN, *Fundamentals of Numerical Reservoir Simulation*, Elsevier, New York, 1977.

16. O.A. PEDROSA, JR., *Use of Hybrid Grid in Reservoir Simulation*, Ph.D. Thesis, Stanford University, California, December, 1984.

17. O.A. PEDROSA and K. AZIZ, *Use of hybrid grid in reservoir simulation*, paper SPE 13507, presented at the SPE 1985 Middle East Technical Conference, Bahrain, March 1985.

CHAPTER 3

An Unknown Ingredient Inverse Problem Reduced to a Functional Differential Equation*

J. R. Cannon**
Paul DuChateau†
Saeed Moaveni‡

Abstract: An inverse problem for the indirect determination of a velocity distribution is formulated and then transformed to a problem for a functional differential equation. The latter problem is shown to be equivalent to the inverse problem. The functional differential equation problem is shown to have a unique solution which can be constructed as the fixed point of a suitable mapping. Results of numerical experiments then illustrate the method.

Introduction. The process of developing a mathematical model for a physical system amounts to finding a mathematical problem whose solution allows us to predict the response of the system to prescribed inputs. Here we will refer to this model problem as the direct problem. As the level of sophistication of mathematical models increases it often happens that some information required as input for a model cannot be obtained by direct measurement. An alternative in such a case is to obtain the information indirectly as the solution of an appropriate inverse problem. This is often accomplished by overspecifying the direct problem and extracting from the overspecified problem a subproblem in which the missing ingredient is the unknown.

Unknown ingredient inverse problems have been treated by a variety of ad-hoc methods over a period of many years (see references [1]-[10]) but recent results indicate the possibility of a systematic approach to the formulation and solution of a certain class of inverse problems. References

*Supported in part by NSF Grant DMS 8901301 and by ONR Contract Number N00014-88-K-0224.
**Department of Mathematics, Lamar University, Beaumont, Texas.
†Department of Mathematics, Colorado State University, Fort Collins, Colorado.
‡Mechanical and Aerospace Engineering Department, Syracuse University, Syracuse, New York.

[11] to [14] illustrate how unknown ingredient inverse problems in which the missing ingredient is a function of a single variable can be transformed to boundary value or initial-boundary value problems for functional differential equations (FDE's). Solution of the FDE problem then is shown to lead to a solution of the inverse problem.

In this paper we are going to illustrate the FDE technique for a typical unknown ingredient inverse problem. We consider the problem of determining the velocity distribution for a fluid flowing in a porous medium; specifically we consider a laminar internal forced convection flow in a rectangular channel. In this situation direct measurements of fluid point velocities are impractical due to the interfering nature of the velocity probes and the degree of difficulty associated with the measurement techniques. Alternatively, for this type of flow a temperature field may be measured more easily and with great accuracy. The following equation can be shown to describe the interaction of the temperature and the velocity fields in this flow situation

$$U(y) \, \partial_x T(x,y) - \partial_{yy} T(x,y) = 0 \qquad (1.1)$$

The coordinate axes have been arranged so that the x-axis lies along the channel in the direction of the flow and the y-axis extends across the channel. Then $T(x,y)$ denotes the temperature field in the fluid and $U(y)$ represents the unknown velocity distribution across the channel. Note that $U = U(y)$ is a function of y only. It may be of interest to note that equation (1.1) also describes 1-dimensional heat conduction in a material with a y-dependent heat capacity function.

We suppose the channel is of unit width, $0 < y < 1$, and we impose the following temperatures on the channel walls

$$T(x,0) = T_0 \qquad \text{and} \qquad T(x,1) = T_1 \qquad (1.2)$$

Finally, we measure the cross-channel temperature distribution at positions $x = 0$ and $x = L$ for some fixed $L > 0$

$$T(0,y) = P(y) \qquad \text{and} \qquad T(L,y) = Q(y) \qquad (1.3)$$

We can now formulate an inverse problem for determining the unknown coefficient $U = U(y)$ in equation (1.1); i.e., for given constants T_0, T_1 and functions $P(y), Q(y)$ find functions $U = U(y)$ and $T = T(x,y)$ satisfying all of the conditions (1.1), (1.2) and (1.3). Note that if the unknown ingredient $U = U(y)$ were known then the problem of finding $T = T(x,y)$ satisfying (1.1) (1.2) and (1.3) is overdetermined. It is our aim to show that when $U(y)$ is not known, then the function pair $\{U, T\}$ is uniquely determined by these conditions.

If $T(x,y)$ satifies (1.1) and (1.2), then the function

$$T(x,y) - T_0(1-y) - T_1 y$$

solves (1.1) as well as homogeneous boundary conditions of the form (1.2). Then there is no loss in generality in supposing that $T_0 = T_1 = 0$ in what is to follow. Now we state several definitions:

The function $P = P(y)$ belongs to the class D of admissible data if

   i) $P \varepsilon \mathbb{C}[0,1] \cap \mathbb{C}^2(0,1)$
   ii) $P(0) = P(1) = 0$ (1.4)
   iii) $P(y) > 0$ for $0 < y < 1$
   iv) $P''(y) < 0$ for $0 < y < 1$

The conditions (1.4)iii and (1.4)iv are convenient for our analysis and at the same time they are conditions that can be arranged experimentally.

The function $U = U(y)$ belongs to the class A of admissible coefficients if

   i) $U \varepsilon \mathbb{C}[0,1]$
   ii) $U(0) = U(1) = 0$ (1.5)
   iii) $U(y) > 0$ for $0 < y < 1$

Note that A is a convex subset of $\mathbb{C}[0,1]$ and of $L^2[0,1]$.

For U in A and P in D given, the function $T = T(x,y)$ belongs to the class S of solutions for the direct problem if

   i) $T \varepsilon \mathbb{C}(\overline{\Omega}_L) \cap \mathbb{C}^{1,2}(\Omega_L)$
   ii) $U(y) \partial_x T(x,y) - \partial_{yy} T(x,y) = 0$ in Ω_L
   iii) $T(x,0) = T(x,1) = 0$ for $0 < x < L$ (1.6)
   iv) $T(0,y) = P(y)$ for $0 < y < 1$

where $\Omega_L = \{0 < x < L, \ 0 < y < 1\}$. Alternatively we could consider the class of coefficients U in $L^2[0,1]$ and correspondingly weaken the sense of the solution to the direct problem. For our purposes we prefer to proceed with A and S as defined.

For P and Q given in D, the pair of functions $\{U, T\}$ is said to form a solution of the inverse problem if

   i) $U \varepsilon A$
   ii) $T \varepsilon S$ (1.7)
   iii) $T(L,y) = Q(y)$ for $0 < y < 1$

The condition (1.7)iii can be viewed as the overspecification in this inverse problem since if U in A is known and T belongs to S then $T(L,y)$ is determined and cannot be independently specified as in (1.7)iii.

We are going to show that there cannot be two function pairs that satisfy (1.7) hence the data in the problem is sufficient to uniquely determine the solution of the inverse problem. In addition we will present a method for approximating the solution to the inverse problem if we assume it to exist. We begin with some auxiliary results.

Auxiliary Results. From [15] we can extract the result that for each P in D and every U in A there exists a unique $T = T(x,y)$ in S. Moreover, this solution T can be represented as follows

$$T(x,y) = \sum_{n=1}^{\infty} P_n \, e^{\mu_n x} \, \varphi_n(y) = e^{xD}\big(P(y)\big) \tag{2.1}$$

where, for $n = 1, 2, \ldots$

$$\varphi_n''(y) = \mu_n U(y) \varphi_n(y), \quad 0 < y < 1, \quad \varphi_n(0) = \varphi_n(1) = 0 \tag{2.2}$$

and

$$P_n = \int_0^1 P(y) \, \varphi_n(y) \, U(y) \, dy \tag{2.3}$$

We are using the notation e^{xD} to indicate the semigroup of solution operators associated with the solution class S. This is an analytic semigroup of self adjoint operators on the weighted Hilbert space $L^2[0,1:U]$. That is for arbitrary P and Q in D we define the weighted inner product by

$$\big(P, Q\big)_U = \int_0^1 P(y) \, Q(y) \, U(y) \, dy$$

Then for x, ξ in $(0,L)$

$$\big(e^{(x+\xi)D} P, Q \big)_U = \big(e^{xD} e^{\xi D} P, Q \big)_U = \big(e^{\xi D} P, e^{xD} Q \big)_U \tag{2.4}$$

Note that the eigenvalues μ_n satisfy

$$\mu_n = \frac{- \int_0^1 \varphi_n'(y)^2 \, dy}{\int_0^1 \varphi_n(y)^2 \, U(y) \, dy} < 0$$

i.e.,

$$0 > \mu_1 > \mu_2 > \cdots \longrightarrow -\infty \tag{2.5}$$

Finally, note that (2.1) implies that for $N = 1, 2, \ldots$

$$\big\| T(x, \cdot) \big\|_U \geq e^{\mu_N x} \big\| \Pi_N(P) \big\|_U \quad \text{for } 0 < x < L \tag{2.6}$$

where Π_N denotes the projection into the subspace spanned by the eigenfunctions φ_1 through φ_N and

$$\left\| \Pi_N(P) \right\|_U = \left(\sum_{n=1}^{N} \left| P_n \right|^2 \right)^{1/2}$$

for P_n given by (2.3).

LEMMA 2.1 For U in A and P in D, let T denote the corresponding unique T in S. Then

$$\partial_x T(x,y) < 0 \quad \text{on} \quad \Omega_L . \tag{2.7}$$

Then for each $L > 0$, $T(L,y)$ belongs to D with

$$T(L,y) < P(y) \quad \text{for} \quad 0 < y < 1. \tag{2.8}$$

Proof- Let $\vartheta(x,y) \equiv \partial_x T(x,y)$. Then $\vartheta(0,y) = \partial_x T(0,y)$ and thus

$$U(y)\,\vartheta(0,y) = \partial_{yy} T(0,y) = P''(y) < 0.$$

Then $\vartheta(x,y)$ satisfies:
$$\begin{aligned}
U(y)\,\partial_x \vartheta(x,y) &= \partial_{yy}\vartheta(x,y) & &\text{in } \Omega_L \\
\vartheta(0,y) &< 0 & &0 < y < 1 \\
\vartheta(x,0) &= \vartheta(x,1) = 0 & &0 < x < L,
\end{aligned}$$

and it follows from the maximum principle that $\vartheta(x,y) = \partial_x T(x,y) < 0$ on Ω_L. This is (2.7) which implies in turn that $\partial_{yy}T(x,y) < 0$ on Ω_L. In particular $\partial_{yy}T(L,y) < 0$ for $0 < y < 1$. Then $T(L,y)$ is in D since T is in S and (2.8) is a direct result of (2.7).

Note that if $\{U,T\}$ is a solution of the inverse problem in the sense of (1.7) then (2.8) implies that the data P and Q must satisfy $Q(y) < P(y)$ for $0 < y < 1$. Thus, in addition to belonging to D, the data must also satisfy $Q < P$ on $(0,1)$.

The Inverse Problem We begin by proving the inverse problem cannot have two solutions. This implies that specifying the extra condition $T(L,y) = Q(y)$, in addition to the usual initial and boundary data for the parabolic problem, is sufficient to uniquely determine the unknown coefficient $U(y)$.

THEOREM 3.1 For data P and Q in D with $Q < P$ on $(0,1)$ and for functions U and V in A, let $T = T(x,y)$ and $\vartheta = \vartheta(x,y)$ in S be such that the function pairs $\{U,T\}$ and $\{V,\vartheta\}$ each solve the inverse problem in the sense of (1.7). Then $U = V$.

Proof- Suppose for some function $p(y)$ in $\mathbb{C}[0,1]$, that $w = w(x,y)$ solves the backward problem,

$$U(y) \, \partial_x w(x,y) + \partial_{yy} w(x,y) = 0 \qquad \text{in } \Omega_L$$

$$\begin{aligned} w(L,y) &= e^{LD}\big(p(y) \big) & 0 < y < 1, \\ w(x,0) &= w(x,1) = 0 & 0 < x < L. \end{aligned} \qquad (3.1)$$

Then,

$$w(x,y) = e^{-xD}\Big(e^{LD}\big(p(y) \big) \Big) = e^{(L-x)D}\big(p(y) \big).$$

The hypotheses of the theorem imply that

$$U(y) \, \partial_x(T-\vartheta) - \partial_{yy}(T-\vartheta) = (U-V) \, \partial_x \vartheta \qquad (3.2)$$

and thus

$$\iint_{\Omega_L} w \Big(U(y) \, \partial_x(T-\vartheta) - \partial_{yy}(T-\vartheta) \Big) dx \, dy = \iint_{\Omega_L} w \, (U-V) \, \partial_x \vartheta \, dx \, dy$$

But

$$\iint_{\Omega_L} w \Big(U(y) \, \partial_x(T-\vartheta) - \partial_{yy}(T-\vartheta) \Big) dx \, dy =$$

$$\iint_{\Omega_L} \big(U \, \partial_x w + \partial_{yy} w \big)(T-\vartheta) dx \, dy = 0$$

and thus, by the mean value theorem for integrals, there exists a $\sigma = \sigma(y)$, $0 < \sigma < L$, such that

$$0 = \iint_{\Omega_L} w \, (U-V) \, \partial_x \vartheta \, dx \, dy = \int_0^1 w(\sigma,y)\big(U(y) - V(y) \big) \int_0^L \partial_x \vartheta(x,y) \, dx \, dy$$

Since ϑ satisfies all of the conditions of (1.7) this last equation reduces to

$$\int_0^1 w(\sigma,y)\big(U(y) - V(y) \big)\big(Q(y) - P(y) \big) dy = 0.$$

Now choose the function $p(y)$ in (3.1) as follows

$$p(y) = \big(U(y) - V(y) \big)\big(Q(y) - P(y) \big).$$

Then

$$w(x,y) = e^{(L-x)D} \big(U(y) - V(y) \big)\big(Q(y) - P(y) \big)$$

$$= e^{2sD} \big(U(y) - V(y) \big)\big(Q(y) - P(y) \big)$$

where $2s = L - \sigma \geq 0$; i.e., $0 \leq s(y) \leq L/2$. Then we have

$$\big(e^{2sD}(p), p \big)_U = \big(e^{sD} p, e^{sD} p \big)_U = \big\| e^{sD} p \big\|^2 = 0$$

But then (2.6) implies that for $N = 1, 2, \ldots$

$$0 = \left\| e^{sD} p \right\| \geq e^{\mu_N L/2} \left\| \Pi_N(p) \right\|_U$$

Then $\left\| \Pi_N(p) \right\|_U = 0$ for $N = 1, 2, \ldots$ which is to say

$$\left\| p \right\|_U = \left\| (U - V)(Q - P) \right\|_U = 0.$$

But $Q < P$ on $(0,1)$ and thus $\left\| U - V \right\|_U = 0$. Finally, since P, Q, U, V are all continuous on $(0,1)$ it follows that $U = V$ on $[0,1]$. Using $U = V$ in (3.2) then leads to $T = \vartheta$ as well, completing the proof.

Note that if we enlarge the class A of admissible coefficients to include bounded semicontinuous functions then we can suitably weaken the definition of the class S of solutions to the direct initial boundary value problem and definition of a solution for the inverse problem. Then Theorem 3.1 is still valid provided we interpret $U = V$ to mean equality almost everywhere on $(0,1)$.

We now show how to use the overspecified condition to eliminate the unknown coefficient $U(y)$ from the partial differential equation and to reformulate the inverse problem as an initial boundary value problem for a functional differential equation. The term functional differential equation is used to indicate an equation containing a coefficient that is functionally dependent on the solution.

For P and Q in D with $Q < P$ on $(0,1)$ and for $T = T(x,y)$ in $\mathbb{C}^{1,2}(\Omega_L)$, define

$$m[T](y) = \frac{\int_0^L \partial_{yy} T(x,y)\, dx}{Q(y) - P(y)} \tag{3.3}$$

Then we have

LEMMA 3.2 Suppose P and Q belong to D with $Q < P$ on $(0,1)$ and that $T = T(x,y)$ is in $\mathbb{C}^{1,2}(\Omega_L)$. If $m[T](y)$ given by (3.3) is in A and

$$\begin{aligned}
m[T](y)\, \partial_x T(x,y) &= \partial_{yy} T(x,y) & &\text{in } \Omega_L \\
T(0,y) &= P(y) & &0 < y < 1, \\
T(x,0) &= T(x,1) = 0 & &0 < x < L
\end{aligned} \tag{3.4}$$

then $T(L,y) = Q(y)$ and the pair $\{ m[T], T \}$ solves the inverse problem.

Proof– The partial differential equation in (3.4) contains the coefficient $m[T]$ which is functionally dependent on the unknown function $T(x,y)$. It follows from this equation that

$$m[T](y) \int_0^L \partial_x T(x,y)\, dx = \int_0^L \partial_{yy} T(x,y)\, dx$$

Since $m[T]$ is in A, lemma 2.1 implies

$$\int_0^L \partial_{yy} T(x,y)\, dx < 0 \quad \text{for } 0 < y < 1.$$

Then

$$\int_0^L \partial_x T(x,y)\, dx = Q(y) - P(y)$$

and since $T(x,y)$ satisfies (3.4), it follows that $T(L,y) = Q(y)$ and that the function pair $\{m[T], T\}$ solves the inverse problem.

For the remainder of this section P and Q will denote a fixed data pair in D with $Q(y) < P(y)$ on $(0,1)$. Then for U in A let $T = T(x,y;U)$ denote the corresponding T in S. We write $T(x,y;U)$ to emphasize the dependence of the solution T on the coefficient U. For this T we have from (3.3) and (1.6) that

$$m[T](y) = \frac{U(y)(T(L,y;U) - P(y))}{Q(y) - P(y)} \qquad (3.5)$$

Now we define a new mapping β by

$$\beta[U](y) = \frac{U(y)(T(L,y;U) - P(y))}{Q(y) - P(y)} \qquad (3.6)$$

and prove

LEMMA 3.3 β maps A into itself
Proof- For arbitrary U in A, $T = T(x,y;U)$ belongs to S and (1.6)i, ii implies that $\beta[U]$ is continuous on $[0,1]$. In addition, (1.6)iii implies $\beta[U]$ vanishes at $y = 0, 1$. Finally, (2.5) implies $\beta[U] > 0$ on $(0,1)$ and thus $\beta[U]$ belongs to A.

Note that $\beta[U] = m[T(x,y;U)]$ and thus if U in A is a fixed point of the mapping β then it follows from Lemma 3.2 that $T(L,y;U) = Q(y)$. In this case the function pair $\{U, T(x,y;U)\}$ solves the inverse problem hence solving the inverse problem can be reduced to the problem of finding a fixed point for the mapping β.

LEMMA 3.4 For U and V in A

$$U > V \text{ on } (0,1) \text{ implies } T(x,y;U) > T(x,y;V) \text{ on } \Omega_L \qquad (3.7)$$

Proof- Let $w(x,y) = T(x,y;U) - T(x,y;V)$. Then by Lemma 2.2

$$U(y)\partial_x w(x,y) - \partial_{yy} w(x,y) = (V(y) - U(y))\partial_x(x,y;V) > 0 \quad \text{on } \Omega_L$$

Since $w(x,y)$ vanishes on the parabolic boundary of Ω_L the maximum principle implies that $w(x,y) > 0$ on Ω_L, proving the result.

Lemma 3.4 implies that the mapping β can be written as the sum of two monotone operators, one isotone and the other antitone. That is

$$\beta[U] = \frac{P(y)}{P(y) - Q(y)} U(y) + U(y) \frac{T(L,y;U)}{Q(y) - P(y)}$$

$$= \beta_1[U] + \beta_2[U] \qquad (3.8)$$

Note that

$$\beta_1[U] - \beta_1[V] = \frac{P(y)}{P(y) - Q(y)} \Big(U(y) - V(y) \Big)$$

and

$$\beta_2[U] - \beta_2[V] = \frac{T(L,y;U)}{Q(y) - P(y)} \Big(U(y) - V(y) \Big)$$

$$+ \frac{V(y)}{Q(y) - P(y)} \Big(T(L,y;U) - T(L,y;V) \Big)$$

Then $Q < P$ and Lemma 3.4 combine to imply that for $U > V$ we have

$$\beta_1 U > \beta_1 V \qquad \text{and} \qquad \beta_2 U < \beta_2 V \quad \text{on } (0,1);$$

i.e., β_1 is isotone and β_2 is antitone.

Each of these mappings is continuous from the convex set A into itself (note that A is a subset of the partially ordered Banach space $\mathbb{C}[0,1]$). Starting with V_0, W_0 in A we define an iteration scheme by

$$V_{n+1} = \beta_1 V_n + \beta_2 W_n$$
$$\text{for } n = 0, 1, \ldots \qquad (3.9)$$
$$W_{n+1} = \beta_1 W_n + \beta_2 V_n$$

Then if a-priori upper and lower bounds for the coefficient U are available, we can show that

$$V_0 \leq V_1 \leq W_1 \leq W_0 \quad \text{on} \quad (0,1) \qquad (3.10)$$

It follows from (3.9) and (3.10) by induction on n that β maps the set

$$M_n = \{ U \in A : V_n \leq U \leq W_n \}$$

into itself (see [16] p353). Then we have

$$V_0 \leq V_1 \leq \cdots \leq V_n \leq V_{n+1} \leq \cdots \leq W_{n+1} \leq W_n \leq \cdots \leq W_1 \leq W_0.$$

The increasing sequence of continuous functions, $\{V_n\}$ is bounded above by W_0 and thus tends to the upper semicontinuous limit $U^* = \sup\{n: V_n\}$. The decreasing sequence of continuous functions $\{W_n\}$ is bounded below by V_0 and so tends to a lower semicontinuous limit $U_* = \inf\{n: W_n\}$. These limits can be shown to be equal almost everywhere on $(0,1)$. The initial boundary value problem has corresponding solutions $T^* = T(x,y; U^*)$ and $T_* = T(x,y; U_*)$ each of which can be shown to satisfy $T(L,y) = Q(y)$ almost everywhere in $(0,1)$. Then either of the U's together with the corresponding solution T for the initial boundary value problem provide a solution for the inverse problem. Note that the solution to the inverse problem is unique in the wider sense described in the remark following the proof of Theorem 3.1.

There are various ways in which a-priori bounds for U in A might be obtained. For example, suppose P in D satisfies conditions which are sufficient to imply that the corresponding solution T in S must satisfy

$$\partial_{xyy} T(x,y) < 0 \quad \text{in } \Omega_L.$$

Then we may conclude that

$$P''(y)\,L \geq \int_0^L \partial_{yy} T(x,y)\,dx \geq Q''(y)L.$$

This leads immediately via (3.5) to the following bounds for U in A

$$\frac{Q''(y)L}{Q(y) - P(y)} \leq U(y) \leq \frac{P''(y)L}{Q(y) - P(y)}$$

Unfortunately it is not readily apparent what conditions on $P(y)$ are sufficient to ensure that $\partial_{xyy} T(x,y)$ is negative in Ω_L. Finding suitable bounds for U in A remains an open problem.

We describe now a numerical procedure for constructing an approximate solution to the inverse problem. For fixed positive integer N, define a uniform partition of $[0,1]$, $\{0 = y_0 < y_1 < \cdots < y_N < y_{N+1} = 1\}$. In addition, define a family of continuous, piecewise linear functions on $[0,1]$

$$b_n(y) = \begin{cases} (y - y_{n-1})/h & y_{n-1} < y < y_n \\ (y_{n+1} - y)/h & y_n < y < y_{n+1} \qquad n = 1, \dots, N \\ 0 & \text{otherwise} \end{cases}$$

These are just the piecewise linear "hat" functions of approximation theory.

For U in A let $u_n = U(y_n)$ for $n = 1, \dots, N$ and note that U in A implies that $u_n > 0$ for each n. Next, define the following projection from A onto $\mathbb{R}_+^N$ the positive half space in $\mathbb{R}^N$

$$\Pi_N U(y) = \{u_1, \ldots, u_N\} \tag{3.11}$$

The mapping
$$\Gamma_N\{u_1, \ldots, u_N\} = \sum_{n=1}^{N} u_n b_n(y) \tag{3.12}$$

goes from $\mathbb{R}_+^N$ into A and the composition $\Gamma_N \cdot \Pi_N$ maps A onto the set

$$A_N = \left\{ U = \sum u_n b_n(y) : u_n > 0 \text{ for } n = 1, \ldots, N \right\} \tag{3.13}$$

i.e. A_N is a convex subset of the finite dimensional subspace spanned by the family $b_1, \ldots, b_N$.

Let P and Q be given in D with $Q < P$ and let

$$p_n = P(y_n), \quad q_n = Q(y_n) \quad n = 1, \ldots, N \tag{3.14}$$

Then given an array $\underline{U} = \{u_1, \ldots, u_N\}$ in $\mathbb{R}_+^N$, we can solve

$$\Gamma_N \underline{U}(y) \, \partial_x T(x,y) = \partial_{yy} T(x,y)$$

$$T(0,y) = P(y) \tag{3.15}$$

$$T(x,0) = T(x,1) = 0$$

for $T = T(x,y;\underline{U})$. Next we compute

$$\hat{u}_n = u_n \frac{p_n - T(L,y_n;\underline{U})}{p_n - q_n} \qquad n = 1, \ldots, N \tag{3.16}$$

and define

$$\{\hat{u}_1, \ldots, \hat{u}_N\} = \beta_N\{u_1, \ldots, u_N\} \tag{3.17}$$

Clearly β_N maps $\mathbb{R}_+^N$ into itself. Moreover, using reasoning analogous to that which established Lemma 3.4, we can show that β_N is the sum of an isotone and an antitone mapping from $\mathbb{R}_+^N$ to itself. We define an iteration scheme based on β_N that is similar to (3.9). Then for functions V_0, V_1, W_0 and W_1 in A such that (3.10) holds, we use the projection Π_N to generate N-tuples $\underline{V}_0, \underline{V}_1, \underline{W}_0$ and $\underline{W}_1$ such that

$$\underline{V}_0 \leq \underline{V}_1 \leq \underline{W}_1 \leq \underline{W}_0 \tag{3.18}$$

Then we can show that for $m = 0, 1, \ldots$ β_N maps the compact, convex set $J_m = [\underline{V}_m, \underline{W}_m]$ into itself. It follows that β_N has a fixed point $\underline{U}^*$ such that

$$\underline{V}_m \leq \underline{U}^* \leq \underline{W}_m \quad \text{for } m = 0, 1, \ldots \tag{3.19}$$

Note that $\Gamma_N \underline{U}^*$ belongs to A_N and that

$$0 \leq \Gamma_N \underline{U}^* \leq W_0(y) \quad 0 \leq y \leq 1. \tag{3.20}$$

The solution of (3.15), $T_N(x,y) = T(x,y;\underline{U}^*)$, satisfies

$$T_N(L,y_n) = q_n \quad n = 1, \ldots, N \tag{3.21}$$

We say that $T(x,y;\underline{U}^*)$ together with $\underline{U}^*$ provide an approximate solution to the inverse problem. In the next section we shall implement this procedure numerically in order to construct these approximate solutions for some specific examples.

Note that (3.18) is critical to the success of this approach and that (3.18) is dependent on (3.10). The condition (3.10) is, in turn, dependent on existence of suitable a-priori bounds for U. Thus a-priori bounds for the unknown coefficients U are needed even for the construction of approximate solution to the inverse problem.

Numerical Experiments. For fixed positive integer N, let $h = 1/(N+1)$ and let $y_n = nh$ for $n = 0, 1, \ldots, N+1$. Let the N-tuples $\{u_1, \ldots, u_N\}$ and $\{p_1, \ldots, p_N\}$ be prescribed respectively by (3.11) for some U in A and by (3.14) for some P in D. Then for positive integer M, the numbers T_{jk} are uniquely determined by the conditions

for $k = 1, \ldots, N$ and $j = 1, \ldots, M$

$$u_k \left(T_{j,k} - T_{j-1,k} \right) = \left(L/Mh^2 \right)\left(T_{j,k-1} - 2\,T_{j,k} + T_{j,k+1} \right)$$

$$T_{0,k} = p_k \quad \text{for } k = 1, \ldots, N \tag{4.1}$$

$$T_{j,0} = T_{j,N+1} = 0 \quad \text{for } j = 1, \ldots, M.$$

Here the positive integer M is chosen sufficiently large that $L/Mh^2 < 1$. Then the equations in (4.1) form an unconditionally stable implicit difference scheme approximating the initial boundary value problem (1.6)ii, iii, iv.

For N-tuples $\{u_1, \ldots, u_N\}$ and $\{p_1, \ldots, p_N\}$ generated by U in A and P in D, we can solve (4.1) to obtain an approximation to $T(x,y;U)$. Using this approximation, we can compute an approximation to $T(L,y;U) = Q(y)$; i.e., $T_{M,k} = q_k$. This numerical solution of th direct problem approximates the physical experiment in which the data Q(y) would be collected. We now submit this artificial data to the solution procedure for the inverse problem.

Using N-tuples $\{p_k\}$ and $\{q_k\}$ as data, we implement the iteration procedure described in the previous section to compute the fixed point $\underline{U}^*$

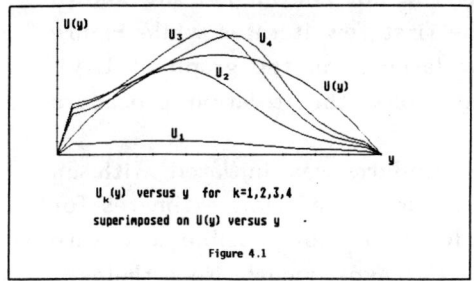

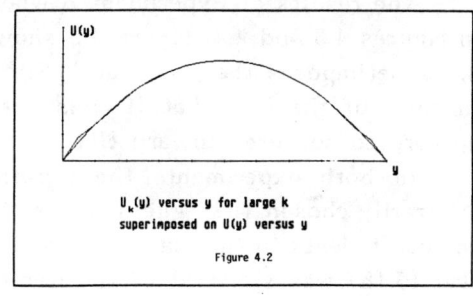

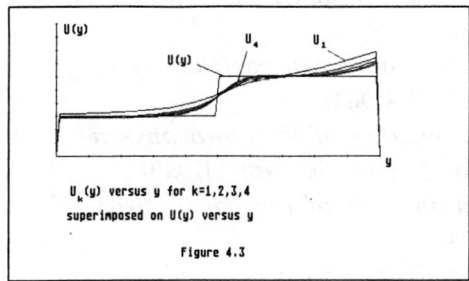

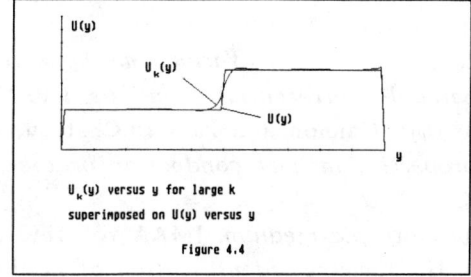

and subsequently obtain the function $\Gamma_N \underline{U}^*$. This process was carried out with $N = 20$, $M = 100$ and $L = 0.1$ for the following examples:

Example A $U(y) = 3y(1-y)$ and $P(y) = 2y(1-y^2)$ $0 < y < 1$.

Example B $P(y) = 2y(1-y^2)$ and $U(y)$ is the semicontinuous function

$$U(y) = \begin{cases} .1 & \text{for} \ \ 0 < y \leq .5 \\ .7 & \text{for} \ \ .5 \leq y \leq 1 \end{cases}$$

Note that $P(y)$ belongs to D and that the $U(y)$ in Example A is in the class A. In each case, the data $Q(y)$ was generated and we then attempted to recover $U(y)$ using (3.16) to generate iterates $\underline{U}^{(k)}$ from an arbitrary initial guess.

The results of Experiment A are shown in Figures 4.1 and 4.2. Figure 4.1 shows the graphs of $\Gamma_N \underline{U}^{(k)}$ for $k = 1, 2, 3, 4$ and Figure 4.2 shows the graph of the iterate for $k = 25$ superimposed on the graph of $U(y)$.

The results of Experiment B where $U(y)$ is not continuous, are shown in Figures 4.3 and 4.4. Figure 4.3 shows the first few iterates while Figure 4.4 superimposes the graph of $\Gamma_N \underline{U}^{(k)}$ for large k on the graph of $U(y)$. In spite of the fact that $U(y)$ is not continuous, the iteration procedure appears to converge toward $U(y)$.

In both experiments the iteration procedure was initiated with an arbitrarily chosen first guess. In particular, since no a-priori estimates for the coefficients in the class A were available, it was not possible to ensure that (3.18) was satisfied by the iterates in our experiments. Nevertheless, convergence occured in every case, independent of the initial guess.

REFERENCES

1. B. Frank Jones, *Determination of a coefficient in a parabolic differential equation, Part I, existence and uniqueness.* Jour. of Math. and Mech, v 11 (6) (1962)
2. _____ , *Various methods for finding unknown coefficients in parabolic differential equations,* CPAM vol XVI (1963)
3. J. R. Cannon and Paul DuChateau, *Determination of unknown physical properties in heat conduction problems,* Int. J. Eng. Sci. vol 11, (1973)
4. _____ , *Determination of the conductivity of an isotropic medium,* JMAA vol 48(3), (1974)
5. U. Hornung, *Identification of nonlinear soil physical parameters from an input-output experiment,* Progress in Scientific Computing, Birkhauser/ Boston, (1983)
6. G. Chavent and P. Lemonnier, *Identification de la nonlinearite d'une equation parabolique quasilineaire,* Appl Math and Opt. vol 1(2) (1974)
7. N. V. Muzylev, *Uniqueness theorems for some converse problems of heat conduction,* USSR Comput Math Phys, vol 20(2) (1980)
8. M. Pilant and W. Rundell, *An inverse problem for a nonlinear parabolic equation,* Comm PDE, vol 11(4) (1986)
9. D. Zachmann, P. DuChateau, and A. Klute, *Calibration of the Richard's flow equation for a draining column by parameter estimation,* SSSA Jour vol 45(6) (1981)
10. P. DuChateau, *Monotonicity and uniqueness results in identifying an unknown coefficient in a nonlinear diffusion equation,* SIAM JAM vol 41(2) 1981
11. J. R. Cannon and P. DuChateau, *An inverse problem for an unknown source term in a heat equation,* JMAA vol 75(2) (1980)
12. _____ , *Weak solutions u(x,t) to a partial differential equation with coefficients that depend on $u(y, \varphi_j(t, u(x,t))$ $j = 1, \ldots, k$* JDE vol 42(3) (1981)
13. _____ , *An inverse problem for an unknown source term in a wave equation,* SIAM JAP vol 45(3) (1983)

14.   and K. Steube, *Unknown ingredient inverse problems and trace type functional differential equations*, Proc of AMS–SIAM Conference on Inverse Problems, 1989 Arcata, Calif.

15. R. E. Showalter, *Degenerate parabolic initial boundary value problems*, JDE vol 31(3), (1979)

16. L. Collatz, *Functional Analysis and Numerical Mathematics*, Acad. Press New York, 1966.

CHAPTER 4

Indicator Evaluation for Self-Adaptive Grid Methods*

J. C. Diaz**
C. G. Macedo, Jr.†
R. E. Ewing‡

Abstract. Many important physical problems exhibit phenomena that require local grid resolution. Activity indicators are used by local grid refinement schemes to determine where to place or remove local grids, in order to achieve a better accuracy on the numerical procedure. The process of indicator selection can be automated by using a procedure which incorporates elements of knowledge-based systems. The methodology requires the creation of a set of model problems representing distinct problem classes, which are used to evaluate the performance of the different activity indicators.

We discuss the application of the methodology to two classes of problems which reveals that the procedure can distinguish correctly between sub-classes within the major classes chosen. One of these classes of problems models transport processes and the other models transport dominated diffusion phenomena. The automated procedure selects different activity indicators for the different classes of problems, depending on the overall characteristics of the localized phenomena present in each of the classes.

Introduction. Many time-dependent processes involve both general and other significantly localized phenomena. These localized phenomena are often critical to the overall chemical and physical behaviors of the processes. For large-scale physical modeling, it is frequently impossible to use a uniform grid sufficiently fine to resolve the local phenomena without yielding a number of unknowns too large for even the largest of supercomputers. Adaptive grid techniques try to overcome this problem by using a combination of grids, see [1,2,3], among others.

---

*Work supported by the National Science Foundation Grants CDA 8820752 and RII-OK-8610676 (Task 10), CNPQ scholarship 20.0269/84, and Oklahoma Center for Advancement of Science and Technology Grants RB9-008 (3748) and ARO-36 (3910).

**Center for Parallel and Scientific Computing, The University of Tulsa, 600 S. College Ave., Tulsa, Oklahoma 74104-3189, U.S.A.

†IBM-Brasil, Brasilia, Brasil.

‡Institute for Scientific Computation, University of Wyoming, P.O. Box 3036, Laramie, Wyoming 82071, U.S.A.

Coarse grids are used for the areas of the domain where localized phenomena are not important. Higher resolution grids are placed in those regions where the process behavior is more dependent on local phenomena. Since the physical processes are often of a dynamic nature, the efficiency of the overall numerical simulation is highly dependent on the ability of the method to dynamically perform grid refinement. High resolution grids have to be created and/or removed, according to necessity, on different regions of the domain.

Adaptive schemes often include three major tasks: grid maintenance, integration, and communication. Grid maintenance determines the placement and/or removal of high resolution grids. Placement or removal of high resolution grids hold the key to the success of the method with respect to the accuracy of the solution. Grid maintenance must assist in keeping track of moving localized phenomena.

One approach to grid placement is to obtain a measure of the error in the actual solution and assume that the integration procedure will generate larger errors at those areas where localized phenomena are part of the solution [1,2,3]. Another way to determine where grid refinement will be necessary is to obtain a measure for an entity that has a direct correlation to the phenomena under consideration. Examples of such entities are the solution gradient or the amount by which the solution differs from neighboring cells. Both of these seem to be attractive methods to keep track of localized phenomena involving discontinuities in the solution. We define an **activity indicator** as any scheme that attempts to identify the occurrence of localized phenomena on a given domain. Activity indicators are used to control the placement or removal of higher resolution grids.

We use an adaptive scheme, originally developed for hyperbolic equations by Berger and Oliger [1],which was modified to also handle parabolic problems. The original scheme used an estimate of the error to control grid placement. The error was estimated using Richardson's extrapolation. It requires two normal integration steps for each grid. The results obtained by twice advancing the solution on the same grid are compared against those obtained using double time step with double cell size in the spatial dimensions. For convection-dominated-diffusion problems, time integration is usually performed using implicit techniques, which require the solution of large sets of linear equations. Further, the refinement process needs to be performed with relative frequency for this class of problems, which makes Richardson's extrapolation costly. It requires two extra integration steps to determine which grids will require the placement of new refined grids, or the removal of existing ones. In other words, since the time spent on integration is directly dependent on the number of grid points, grid refinement analysis using this error estimator imposes a time penalty equal to 125% of the time spent to perform the integration on the grid. The penalty factor would be larger but the method trades increased memory allocation for reduced time by storing the actual solution and the solution at the previous time step.

In trying to determine the optimal indicator for a given problem, and as an alternative to Richardson's extrapolation, different activity indicators are considered, such as gradient, second derivative, percent-change in solution value, actual solution value, etc. The selection of the optimal indicator depends on the method and the performance parameters used to evaluate the indicators. They should be measured with respect to their effectiveness to accurately model both the general and the localized phenomena present on the domain.

Several performance parameters can be considered as candidates for activity indicator evaluation: total time spent in the numerical procedure, magnitude of the error(s) in the final solution, the number of grids created and where they are placed, etc. Since each of these performance parameters quantifies effectiveness of an indicator from a different perspective, results have the tendency to point in different directions, see Díaz and Macedo [5] and Macedo *et al.* [6]

Macedo *et al.* [4] proposed an automated methodology to assist the process of indicator selection for different classes of problems. The automated methodology incorporates concepts from knowledge-based systems, see Rich [7] and Jackson [8]. Here the methodology is applied to two classes of problems revealing that the procedure can distinguish correctly between sub-classes within the major classes chosen. The presentation is organized in four sections. First, the methodology is briefly reviewed. The following sections include the definition of the classes of problems, and the specification of the selection performance parameters. The last section presents some results obtained by applying the automated procedure for the two classes of problems. We finalize by stating our conclusions.

Automated methodology using knowledge-based systems. In general terms, the proposed methodology for the evaluation process consists of four activities. First, a class of problems is selected for analysis and a set of model problems belonging to this class is constructed. Each class should be clearly defined to cover the different types of localized phenomena depicted by its members. Second, a set of performance parameters deemed to be appropriate for the evaluation of the selected activity indicators is constructed. For each performance parameter, a scoring equation is assigned and entries on a database are initialized to reflect this assignment as well as to reflect the fact that the evaluation is about to begin.

On the third step, the automated procedure runs the adaptive grid refinement program, solving each problem in the representative set several times, using each time one of the activity indicators considered. Data is collected for each of the performance parameters involved in the evaluation, and stored in a intermediate data set. At the end of this data collection phase, the performance of each indicator with respect to each performance parameter can be assessed by comparing the data obtained for each indicator against the data for the others indicators, for the same performance parameter. The assessment is done by assigning a score for each indicator with respect to each performance parameter. At the end of this step, each indicator has a set of scores which reflects its performance in the modeling of the class currently under consideration.

A final score for each indicator is constructed by combining the intermediate scores. Each parameter score contributes to the final score with its own weight. The weights are chosen to reflect the importance attached to each performance parameter. After this two-step scoring process, each indicator has been given a score that reflects its performance when used to solve a model problem within the class being studied. These scores are used to update a database using a weighted average between the newly obtained scores and those already existent in the database.

If the number of model problems is sufficiently large, the scores in the database will stabilize and will provide immediate information concerning the quality of each indicator compared with the others. If two or more indicators happen to have similar scores after this process, it could signify that no indicator alone could perform the complete task of signaling the grid refinement or reduction, but that a combination of the most successful indicators might be required.

Classes of problems. The methodology is applied to two very general classes of problems. The first class consists of convection-diffusion problems represented by the parabolic equation

$$u_t = \varepsilon \times (u_{xx} + u_{yy}) + u_x + u_y, \quad in \quad [0, 1] \times [0, 1].$$

Within this class we chose boundary and initial conditions to generate three types of problems.

Problem type 1. It models a front initially located on the vicinity of corner (0, 0) and moving towards the opposite corner (1, 1). This is a typical phenomenon present when modeling the concentration of an invading fluid.

Problem type 2. It models a triangular shaped wedge moving from near (0, 0) towards (1, 1). The base of the triangle is varied from 0.01 to 0.10. The attempt is made to model some of the typical phenomena present in a combustion process where the temperature has a similar profile near the ignition point. Such problems occur in the simulation of combustion modeling of hydrocarbon reservoir exploitation.

Problem type 3. It simulates a truncated wedge; the increase to the plateau value, and the decrease from it, is done with a linear ramp of width d which varies between 0.01 and 0.10, on each side of the plateau. Like the other two problems, it moves from (0, 0) to (1, 1).

All three types of problems are considered with $\varepsilon = 1.0$, 0.1, and 0.01 (making the problems increasingly more transport dominated). Types 2 and 3 had $d = 0.10$, 0.05, and 0.01. A total of 21 different problems are considered. A complete specification of the boundary and initial conditions is given in Macedo *et al.* [6].

The second class consists of transport problems described by one of the following hyperbolic equations. Five different problem cases are considered.

$$u_t = - y \times u_x - x \times u_y, \qquad in \ [0, 1] \ x \ [0, 1], \tag{A}$$

$$u_t = - u_x - u_y, \qquad in \ [0, 1] \ x \ [0, 1], \tag{B}$$

$$u_t = u_x + u_y, \qquad in \ [0, 1] \ x \ [0, 1]. \tag{C}$$

Rotating cone. This well known problem is modeled by equation A. The initial conditions are as follows. Let

$$z(x,y) = (x - 0.25)^2 + 1.5 \times y^2.$$

Then

$$u(x,y,0) = 1.0 - 2.0 \times z(x,y), \qquad if \ z(x,y) < 0.5,$$

$$u(x,y,0) = 0.0, \qquad otherwise,$$

with boundary condition

$$u(x,y,t) = u(x,y,t-1), \qquad for \ t \geq 1,$$

for any point (x,y) on the boundary of the region under consideration.

Expanding wave. This problem is defined by equation B and an initial condition with two different levels of concentration on the domain. It models the dilution in time of a circular region of high concentration over the whole domain. Many chemical transport processes follow this model. The initial conditions are

$$u(x,y,0) = 14.0, \qquad if \ (x-0.5)^2 + (y-0.5)^2 \leq 0.04,$$

$$u(x,y,0) = 1.4, \qquad otherwise,$$

with boundary condition

$$u(x,y,t) = u(x,y,t-1), \qquad for \ t \geq 1,$$

for any point (x,y) on the boundary of the region under consideration.

Moving fronts. The last three problems, chosen as representatives for the hyperbolic class, are defined by equation C which is the same as the equation defining the problems for the parabolic class with $\varepsilon = 0.0$. The first problem in this group, identified as Front 1, uses the same boundary and initial conditions as the problem type 1. The second problem, identified as

Front 2, uses the same boundary and initial conditions as the problem type 2, with the base of the triangular wedge, d, equal to 0.1. Finally, the third problem, identified as Front 3, uses boundary and initial conditions similar to those for problem type 3, with the base of the truncated wedge, d, set to 0.1.

Specification of performance parameters. In this section the indicators considered in the study and the performance parameters used to determined the optimal indicator, are specified. The indicators selected for testing are:

percent of change from closest neighbors (**%Neigh**),
percent of change from previous solution step (**%Prev**),
gradient approximation (**Grad**),
function value (**Value**),
second derivative approximation (**2ndD**), and
Richardson's extrapolation (**RichE**).

Because error estimation using Richardson's extrapolation is considered, from the outset, to be too expensive for problems where regridding occurs with a high frequency, it is not selected as a potential candidate for the parabolic class. The hyperbolic problems are solved with all six activity indicators with the expectation that at least one of the first five indicators would provide solutions as good as those obtained with the Richardson's extrapolation procedure, at a fraction of the cost, in terms of total execution time.

The performance parameters chosen for evaluation of the defined set of activity indicators are:

total number of points in higher resolution grids (**Points**),
total number of grids created (**Grids**),
maximum error in coarsest grid * total execution time (**MaxE**),
average error in coarsest grid * total execution time (**AvgE**),
total execution time (**Time**), and
H^1 error estimate in coarsest grid * total execution time (**H^1**).

The **Points** and **Time** performance parameters are chosen because they provide a measure of how much work is done during the grid refinement process. **Points** differs from **Time** in the sense that it refers to the amount of work done by the integration task, which normally accounts for much of the total execution time, while **Time** evaluates the overhead introduced by grid refinement and communication. The **MaxE**, **AvgE**, and **H^1** performance parameters measure how much effort is necessary to achieve a given degree of accuracy for the modeling procedure since all three can be interpreted as an estimate of the number of accurate digits on the final solution values, per unit of time. The **Grids** parameter measures the potential for parallelism inherent to the process. Each grid on a level is independent from the other grids at the same level. Hence, they could be integrated concurrently when using a multi-processor system. Thus if all other factors are kept constant, it would be more advantageous to have several grids at each refinement level instead of just one per level.

All parameter values, except for **Grids**, have an inverse relationship with the "goodness" of the indicator. The higher their values the worse the indicator performed. To calculate the score assigned to each indicator with respect to a given parameter, a linear scoring equation is used which forced all scores within the range 10 to 100, with 100 given to the best indicator.

Two sets of weights are chosen in order to study the effect of weight selection on the final scores for each indicator, after all the model problems in each class had been solved. With the first set, the emphasis is on the maximum and average errors, while the second set shifted the emphasis to the H^1 estimate of the error. The actual values selected for each set are shown in Table I. Given a weight set, the final score for an indicator, for any given problem, is calculated by summing all its intermediate scores after they are scaled using the corresponding weight. Since the sum of all weights (using set 1 or set 2) equals 1 the final score of an indicator is also kept in the range 10 to 100.

Table I Weights for Performance Parameters

set	Points	Grids	MaxE	AvgE	H^1	Time
1	0.05	0.05	0.35	0.35	0.10	0.10
2	0.05	0.05	0.10	0.10	0.60	0.10

RESULTS

In the parabolic class, problem type 1 had three cases depending on the value of ε and problems type 2 and 3 had nine cases each, depending on the values of ε and the parameter d, the base width of the triangular wedge. Macedo *et al.* [6] presented the scores for the parabolic class by problem case. Table II shows the final scores when these are calculated by problem type.

The final database scores are obtained by combining all the indicator scores for each problem case. It shows that, overall, the solution value indicator is the most appropriate choice given the class of problems being considered, namely, parabolic problems with moving fronts. **Value** is adversely affected by a change on the weights emphasizing the H^1 error estimate, but still has a much better score than the next possible choice, **Grad**. The gradient-based indicator would be the next best choice and its scores do not seem to be significantly influenced by modifications of the weight set values. The other three indicators have scores similar to the score for **Grad** when using weight set 1, but the percent change indicators are adversely affected if the H^1 set is chosen, specially **%Prev**.

For problem type 1, the best indicator is the second derivative, no matter which set of weights was used. However, analysis of the scores by problem case indicates that for problems where the localized phenomena are characterized by a discontinuity in the solution, a high diffusion factor would require the use of the second derivative indicator. Problems where the dominating factor is convection would be better suited for the indicator based on solution values or gradient.

Table II Final scores for parabolic class

Problem type	Weight set	Indicator Scores				
		%Neigh	%Prev	Grad	Value	2ndD
1	1	44.76	56.69	68.99	73.56	82.03
	2	32.29	44.24	59.18	68.88	87.72
2	1	45.80	49.61	65.76	92.47	40.71
	2	44.85	41.10	65.62	92.91	37.79
3	1	56.23	55.47	50.50	71.24	54.81
	2	54.55	45.62	52.05	61.25	54.64

Recall that problem type 2 is a triangular shaped wedge moving across the domain, ideally maintaining its shape and amplitude, given the boundary conditions used. For a small triangular base (d = 0.01), the solution value indicator is clearly superior, with **Grad** and **2ndD** alternating as second choice. For a wider base (d = 0.05) the gradient becomes a good second choice when the value of ε is equal to 1.0; **Grad** becomes the method of choice when the triangular base has the width of the coarse grid space size (d = 0.1), with **Value** now trailing in second. The weight set used did not play a significant role for this problem type,. Overall, the indicator based on solution values outperforms all other possible candidates, with final scores for both weight sets almost 50% higher than the closest competitor, gradient.

Problem type 3 simulates a truncated triangular shaped wedge; on both sides of the plateau, of width d are used making the total width of the wedge equal to $2d$. For this problem there is no clear best choice. For an ε of 0.10, **%Neigh** (percent of change from neighbors) had a couple of perfect scores but failed badly in some other cases. The solution value indicator performed well in most cases but its scores dropped significantly if the H^1 set of weights is used. **2ndD** scored well whenever the value of ε is low, a behavior also shared by the gradient indicator. **%Prev** had a very erratic behavior. For all problem cases, solution values is the indicated choice, followed by **%Neigh** and **2ndD**. Notice that the weight set emphasizing the H^1 error significantly affects **Value**, dropping its final score by more than 10%.

An important observation can be made from the results shown in Table II. The automated procedure is capable of differentiating the problem cases in two sub-categories, invading fronts (as modeled by problem type 1) and moving fronts (as modeled by problem types 2 and 3). The distinction becomes evident when examining the final scores grouped by problem type. One indicator is better for one sub-category while another is more appropriate for the second one.

Table III shows the results obtained for the hyperbolic problems. Again, for each problem type, the first row of scores are obtained using the weight set which emphasizes the average and maximum error on the coarsest grid. The second row of scores are calculated using the weight set which gives greater importance to the error measured in the H^1 norm, also on the grid with the coarsest level of refinement.

Table III Final scores for hyperbolic class

Problem type	Weight set	Indicator Scores					
		%Neigh	%Prev	Grad	Value	2ndD	RichE
Cone	1	71.29	50.07	92.49	92.37	27.00	50.35
	2	81.74	53.06	91.53	94.60	54.42	25.93
Circle	1	67.67	95.50	67.69	37.56	14.83	59.87
	2	80.93	95.50	80.95	38.49	14.59	53.57
Front 1	1	47.91	91.41	73.94	18.08	58.61	74.34
	2	27.42	94.33	62.51	35.96	49.50	72.22
Front 2	1	58.27	87.19	45.35	74.35	40.26	14.85
	2	62.30	92.77	48.89	79.08	26.20	18.62
Front 3	1	13.25	89.62	59.95	79.02	78.86	32.93
	2	13.25	94.01	61.03	69.19	63.86	30.44
Combined	1	51.68	82.76	67.88	60.28	43.91	46.47
	2	53.13	85.93	68.98	63.46	41.71	40.16

The rotating cone problem is effectively solved using either **Grad** or **Value**, with **%Neigh** also performing reasonably well. For this problem the other indicators did a poor job compared to the first three. For the other four problem cases, the clear winner is the indicator based on the percent of change of the solution from one time step to the next, **%Prev**. The **Value** indicator performed well for Fronts 2 and 3, the triangular shaped wave, but is badly penalized by the scores for the problems that simulate an invading front. The gradient indicator had somewhat consistent scores across all problems, with the remaining indicators obtaining low scores for almost all problems.

The overall winner for the class of hyperbolic problems is **%Prev**, followed by **Grad** and **Value**, with the latter two having very close scores. The results again point towards the fact that the problems selected as representatives of the class are not completely similar and really could be grouped in three sub-classes: the rotating cone as an example of a moving mass, invading front problems exemplified by the Circle and Front 1 problems, and moving waves represented by problems Front 2 and 3. The procedure correctly distinguished between the sub-classes by pointing out different optimal indicators for each class. This behavior does not detract from the method's capability of helping to choose the most appropriate indicator for a given class of problems. It suggests that the set of model problems has to be carefully chosen.

As expected, the indicator using Richardson's extrapolation procedure is not selected as the optimal choice for any of the problem cases, confirming the belief that it is costly and can be replaced by other rules as the main control for the task of grid placement and removal, without loss of performance in terms of the final solution results.

Conclusions. The placement and removal of high resolution grids is an important part of an adaptive-grid numerical method for the modeling of problems having both global and localized phenomena as part of the solution. Ultimately, the solution obtained with such numerical methods is as good as the method's ability in tracking the localized phenomena over the domain. To help select the activity indicator which is best suited for a given class of problems, an automated methodology can be used. The automation of the indicator evaluation process is achieved using a knowledge-based procedure. It self-adjusts its parameters as it solves different problem cases, using all the indicators selected for evaluation.

Two very general classes of problems are used as case studies to illustrate the automated methodology. Looking at the overall results for each class it is easy to decide on the best indicator for each class, given the set of performance parameters chosen. If the model problems representing the class can be grouped in sub-classes, the automated procedure differentiates between them by pointing toward distinct indicators as the optimal choices for each sub-class. The final scores for each indicator can be affected by the weights chosen. These are a measure of their relative importance within the evaluation process. This stresses the fact that the results obtained using this methodology are dependent on the choices for criteria, weights, scoring equations, and number and type of problems chosen as representatives of a class.

Because all the relevant information for the execution of the automated methodology is stored as a set of rules in a database, the procedure can be easily modified to utilize different indicators, performance parameters, or scoring equations. Human intervention is important for selecting the performance parameters, model problems, and weight set, but the bulk of the process can now be assigned to a machine.

Acknowledgements. We want to thank the Mathematics and Computer Science Division of Argonne National Laboratory for the use of the Alliant FX/8 in their Advanced Computer Research Facility.

References

1. Berger, M. J., Oliger, J., *Adaptive mesh refinement for hyperbolic partial differential equations*. Man. NA-83-02, Comp. Science Dept., Stanford University, March, 1983.

2. Díaz, J. C., Ewing, R. E., Jones, R. W., McDonald, A. E., von Rosenberg, D. U., Uhler, L. M., *Self-adaptive local grid refinement for time-dependent, two-dimensional simulation, Finite Elements in Fluids, Vol VI*, (eds, Gallagher, et al.), John Wiley and Sons, 1985, pp 279-290.

3. Ewing, R. E., *Adaptive Grid Refinements for Transient Flow Problems, Adaptive Methods for Partial Differential Equations* (eds. J.E. Flaherty, P.J. Paslow, M.S. Shephard, J.D. Vasilakis), SIAM, 1989, pp 194-205.

4. Lucier, B. J., *A stable adaptive numerical scheme for hyperbolic conservation laws.* SIAM J. Numer. Analysis vol 22, 1985, pp 180-203.

5. Díaz, J. C., Macedo, Jr., C. G., *Parameter selection using knowledge-based systems for the numerical solution of PDE's.* Proceedings of the II Oklahoma Symposium on Artificial Intelligence, 3-4 Nov., OCCE, Norman, OK, 1988.

6. Macedo, Jr., C. G., Díaz, J. C., Ewing, R. E., *A Knowledge-based System for the Determination of Activity Indicators for Self-Adaptive Grid Methods.* Math. and Comp. Simulation vol 31, 1989, pp 431-439.

7. Jackson, P. C., *Introduction to Artificial Intelligence.* Petrocelli Books, 1974.

8. Rich, E., *Artificial Intelligence.* McGraw-Hill, 1983.

CHAPTER 5

Markov Chain Length Effects on Optimization in Groundwater Management by Simulated Annealing

David E. Dougherty*
Robert A. Marryott*

1. INTRODUCTION

Optimization methods are being used increasingly, in conjunction with simulation models, to assist in the design of least-cost solutions of groundwater contamination problems [Gorelick *et al.*, 1984; Ahlfeld *et al.*, 1988; Dougherty and Marryott, 1990]. Flow and transport simulators that approximately solve systems of partial differential equations are employed to evaluate the costs and feasibility of candidate solutions. These simulations dominate the computational requirements of simulated annealing and other optimization methods. Therefore, its is advantageous (1) to seek speedier simulators and (2) to identify or develop optimization methods that require a minimal number of simulations so that more comprehensive design alternatives can be explored. We shall focus in this article on the latter approach.

Simulated annealing has recently been applied to optimization problems in groundwater management [Dougherty and Marryott, 1990]. The method, given in Table 1, yields good solutions to problems cast in combinatorial form. Unfortunately, the straightforward implementation of the method has very large computational requirements through its appetite for transport simulations. Several techniques to reduce the number of simulations required can be advanced by careful specification and modification of the various components of the annealing algorithm.

The simulated annealing method requires that five components be described:

1. A concise representation of (data structure for) the system configuration, $\mathbf{r}_i$.

---

*Department of Civil Engineering, University of California, Irvine, CA 92717.

INITIALIZE:
$T = T_0$;
$i = i_0$;
$C = C(i_0)$
DO WHILE (stopping criterion not satisfied);
    DO WHILE (equilibrium not satisfied);
        PERTURB:
        $j = \text{rearrangement}(i)$;
        $\Delta C = C(j) - C(i)$;
        IF $\Delta C \leq 0$ **THEN**
            ACCEPT:
            $C = C(j)$;
            $i = j$;
        ELSE
            IF $\text{random}(0,1) < \exp(-\Delta C/T)$ **THEN**
                ACCEPT:
                $C = C(j)$;
                $i = j$;
            ENDIF
        ENDIF
    ENDDO
    $T = \alpha T$
ENDDO

Table 1: Pseudocode description of the simulated annealing algorithm.

2. A scalar cost function, $C(\mathbf{r}_i)$, which expresses the objectives of the optimization as a single number and also quantifies tradeoff among multiple constraints and/or objectives.

3. A procedure for generating random changes or rearrangements of the system.

4. A control parameter T and an annealing schedule.

5. A criterion for terminating the algorithm.

Item 1 has little effect on the cost of the method, because for most of the applications we consider the simulations dominate the computational effort. Item 5 does have an effect, but because the quantitative statement of this is relatively precise (the variance of cost within a specified number of temperature steps being essentially zero) there is relatively little one can do with this short of premature termination of the algorithm.

Item 2 has an influence on the work required because the cost function describes the degree to which the cost is coupled to the modules of the flow and transport simulator. For example, if the cost function includes a cost proportional to the value of the control variable and on the flow variables only—with no contribution from the transport simulation—then the simulation work is greatly reduced. In Dougherty and Marryott [1990] we present a screening method that decomposes the cost function into parts dependent directly on the control variables and implicitly dependent on them. Thus, totally unreasonable alternatives are weeded out without performing expensive simulations.

The procedure for generating system rearrangements, Item 3, has a discernible effect on the number of simulations required and on the rate of convergence of the annealing process [Dougherty and Marryott, 1990]. This is true for other methods of optimization as well; the search strategy in the traveling salesman problem provides a classical example [Lin, 1965]. We have developed a method of restricting searches that depends on the state of the system. This "excursion limiting" method is dependent on the application being tackled by the optimizer. Basically, we limit the size of the neighborhood of a state which can be explored to find better solutions in accordance with the state and the status of the optimization procedure.

The annealing or cooling schedule has a major influence on the quality of the computed optimal solution and the amount of computation employed to obtain it. The annealing schedule comprises the following components:

1. The initial value of the optimization parameter (temperature) T_0.

2. The number of system rearrangements made at each value of the temperature.

3. The decrease in temperature from one step t, in the annealing process to another, $T_{t+1} \leftarrow \alpha_{t+1} T_t$.

The first of these is very important in obtaining nearly optimal solutions yet has very little bearing on the overall cost of the optimization because of the way the temperature is reduced—an excessively high T_0 yields only a few extra values of temperature that must be sampled during the optimization. The third element of the annealing schedule plays an important role as well; if the temperature is reduced at too fast a rate the cost determined as "optimal" will be too high and if the temperature is reduced too slowly the quality of the solution will be excellent but the required computational resources will be excessive. A logarithmic decrement of temperature based on the solutions sampled at a given temperature [Huang *et al.*, 1986] provides an effective control mechanism for temperature decrement, by computing the temperature decrement factor α as

$$\alpha_{t+1} = \max \left\{ \begin{array}{l} \exp(-0.7 \; T_t/\sigma_t) \\ 0.5 \end{array} \right. \tag{1}$$

Here, σ_t is the standard deviation of the costs encountered during the stage with temperature T_t.

The number of configurations of the system investigated at each temperature, i.e., the length of Markov chain at each temperature, is critical, both in terms of the degree of optimality of the solution and the cost of the method. This is the object of our attention in this article.

2. MARKOV CHAIN LENGTH

Aarts and van Laarhoven [1985] showed that if the method of simulated annealing were treated as a homogeneous Markov chain problem then convergence to the globally optimal solution of the optimization problem occurs with probability one as the temperature is reduced to zero. Unfortunately, their analysis put an excessive number of restrictions on the rearrangement method and the annealing schedule, provided no information on the rate of convergence to optimality, and could specify only in the broadest (and impractically large) terms the number of rearrangements needed at each temperature to reach the optimal solution.

Mitra *et al.* [1985] extended the analysis of simulated annealing by more realistically treating it as a nonhomogeneous Markov process. Assuming a particular temperature decrementing rule, this paper again showed convergence to global optimality with probability one as the temperature dropped to zero. Furthermore, an estimate of the rate of convergence to the optimal solution was determined. It was found to depend on the difference between the least and next-least costs of the the system being optimized, and (inversely) upon the rate of temperature reduction.

This result is much more useful. However, the least and next-least costs are generally not known for problems involving groundwater remediation, so little quantitative guidance for the necessary Markov chain length at each temperature level can be extracted.

Kirkpatrick *et al.* [1983], in the article introducing simulated annealing as an optimization method, suggested that the chain length be some factor times the number of decision variables. Huang *et al.* [1986] further recommended that this factor be adjusted during cooling. We have found through experience [Dougherty and Marryott, 1990] that about 100 times the number of decision variables gave high quality optimization results for the problems of interest to us.

We have resorted to computational experiments to develop further guidance on this issue. Groundwater flow and transport examples were employed in a Monte Carlo experiment. For each problem, a number of runs were made at each of several Markov chain lengths. The quality of the solutions was available through the costs for the multiple runs at each chain length.

3. COMPUTATIONAL EXAMPLES

We shall assess the effect of Markov chain length on two problems; one drawn from a groundwater quantity and the other representative of remediation problems for groundwater contamination incidents. In both cases we use up to 100 Monte Carlo runs for each of several chain lengths to determine

1. The cost of the optimal solution obtained for each chain length (C_l where l is a chain length index) compared with the minimum cost solution obtained over all experiments with all chain lengths (C_u).

2. The frequency with which a given chain length leads to costs of C_u.

3. The minimum chain length that will produce acceptable suboptimal solutions, which is expected to vary from one problem type to another.

3.1 Dewatering Problem

We consider the reduction of head at a point in an aquifer caused by pumping from a number of wells, N_w. A linear response function is assumed, so that the drawdown at the point may be expressed as

$$h = \sum_{w=1}^{N_w} F_w Q_w, \tag{2}$$

where h is the drawdown at the point due to all wells, Q_w is the pumping rate from well w, and $F_w Q_w$ is the drawdown at the point of interest due to the pumping at well w. For convenience only we shall assume F_w is a constant for all wells.

The optimization problem of interest is:

$$\text{minimize } C(\mathbf{Q}),$$

where $\mathbf{Q}$ is the vector $\{Q_1, \ldots, Q_W\}$ and

$$C(\mathbf{Q}) = \sum_{w=1}^{N_w} C_{Qw} Q_w + \sum_{w=1}^{N_w} C_{2w} Q_w^2$$

$$+ \sum_{w=1}^{N_w} C_{Iw} N(\mathbf{Q}) + P\gamma(h, \overline{h}). \tag{3}$$

Here $N(\mathbf{Q})$ is a function indicating the number of wells pumping at nonzero rates, C_{Qw} is the cost per unit of pumping at well w, C_{2w} is cost per unit of the square of the pumping rate of well w, C_{Iw} is the cost of installation per pumping well, $\gamma(h, \overline{h})$ is a cost function associated with not achieving the desired value drawdown $\overline{h}$, and P is a scaling factor for inadequate drawdown. Again, for convenience, we shall assume the unit costs are identical for all wells. γ is selected to be

$$\gamma(h, \overline{h}) = \begin{cases} 0, & h \geq \overline{h} \\ (h - \overline{h})^2 & h < \overline{h} \end{cases} \tag{4}$$

Let $N_w = 5$ and allow six possible pumping rates at each well, $\{0, 1, 2, 3, 4, 5\}$. The configuration space contains $6^5 = 7776$ possible states, which is small enough for a complete search to assess the minimum cost. In this example the cost function is very simple and can be computed inexpensively.

Our Monte Carlo tests were carried out for each of 10 sets of cost coefficients; $\overline{h} = 10$ and $F_w = 2$ for all cases. The sets of coefficients are given in Table 2. For each case, Markov chain lengths of 5, 20, 35, 50, 70, and 100 were considered with 100 replicates per chain length and set of cost coefficients. Table 2 also indicates the optimal costs, C_u, determined by slow annealing with very long chain lengths.

Figures 1 and 2 illustrate the results obtained for cases 2 and 6. Panel (a) in each figure shows the average optimal cost for the 100 Monte Carlo runs at each chain length, while panel (b) describes how frequently the ultimate optimal cost of a case, C_u, was obtained for a given chain length. Cases 1 and 10 yielded essentially the same results as case 2, although the costs C_u differed. The behavior shown in Figure 2 typifies cases 5, 6, 7, 8, and 11. Cases 9 and 12 exhibited intermediate behavior. The results show that the chain length needed depends on the coefficients of the cost function, particularly, it appears, on C_2.

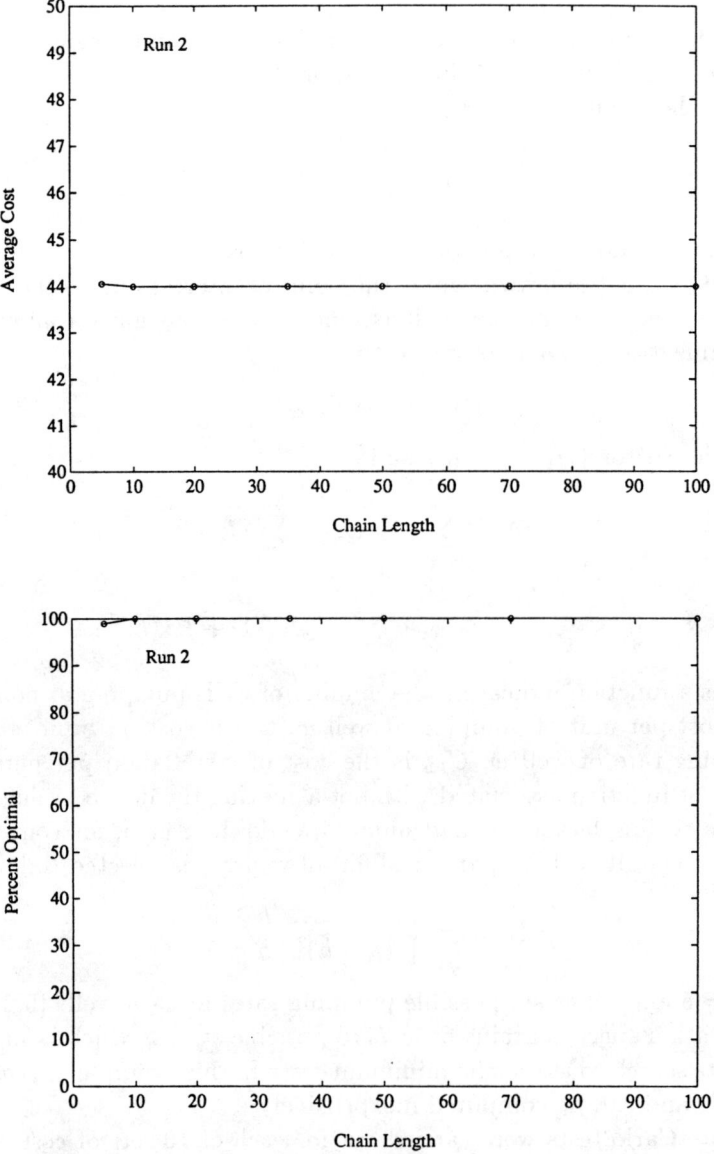

Figure 1: Results of Monte Carlo experiments on drawdown case 2. (a) average costs of optimized solution and (b) percent of solutions attaining actual optimal solution versus Markov chain length.

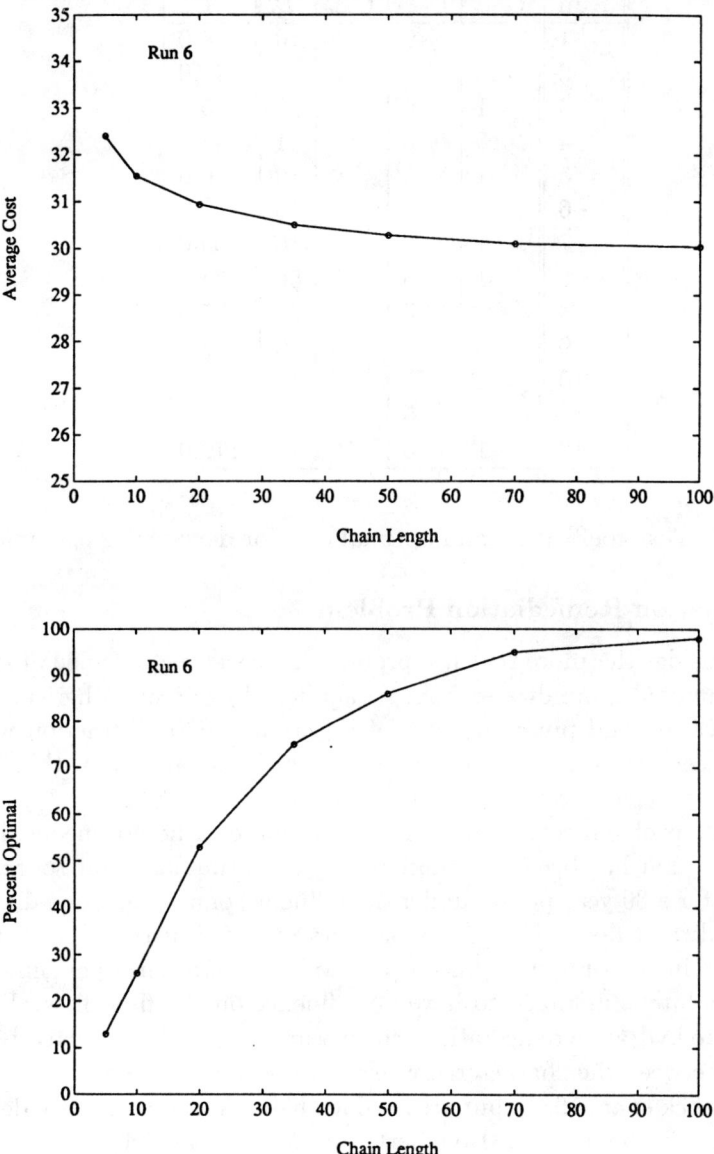

Figure 2: Results of Monte Carlo experiments on drawdown case 6. (a) average costs of optimized solution and (b) percent of solutions attaining actual optimal solution versus Markov chain length.

Run	C_{Qw}	C_{2w}	C_{Iw}	P	C_u
1	1	0	0	10	5.0
2	10	0	0	1	44.0
3	1	0	0	10	5.0
4	10	0	0	1	44.0
5	1	5	0	10	30.0
6	5	1	0	10	30.0
7	1	5	1	10	35.0
8A	1	5	5	50	55.0
8B	1	5	5	50	55.0
9	5	1	5	50	48.0
10	1	5	10	10	80.0
11	1	5	1	50	35.0
12	1	5	50	10	170.0

Table 2: Cost coefficients and optimal cost for dewatering example.

3.2 Contamination Remediation Problem

We now consider the more realistic problem of designing a remedial system to remove contaminated groundwater from an aquifer. In this hypothetical example, the optimal location and pumping rates of injection and/or extraction wells that will reduce concentration levels below a given threshold value, $\bar{c}$, within a certain time-frame in a specified area must be determined.

The example problem represents the remediation of a homogeneous, isotropic, confined aquifer that has been contaminated by a continuous point source of non-reactive solute for a 30 year period under the influence of a steady, one-dimensional regional groundwater flow. The plume of contaminant, illustrated with the computational domain Ω in Figure 3, has a peak value of 115 parts per billion before remediation. Solute is assumed to have no influence on the fluid flow. Fluid flow transients due to hydraulic remediation schemes are frequently neglected because of the difference between the characteristic relaxation times of the flow and transport equations [Gorelick *et al.*, 1984], but are retained here. A more complete description of this example may be found in Dougherty and Marryott [1990].

The objective of the remediation design is to minimize the cost

$$C = \sum_{w=1}^{N_w^+} C_{Q_w^+} Q_w^+ + \sum_{w=1}^{N_w^-} C_{Q_w^-} Q_w^- + \sum_{w=1}^{N_w} C_{I_w} N(\mathbf{Q}) + P\gamma(c,\bar{c}) \qquad (5)$$

where N_w is the number of wells ('+' for injection, '−' for extraction), Q_w is the pumping rate at well w, $C_{Q_w^+}$ is the cost factor per unit rate of injection, $C_{Q_w^-}$ is the cost factor per unit rate of extraction, C_{I_w} is the cost of installation for well w, $N(\mathbf{Q})$ is an indicator of wells pumping at nonzero rates, P is the penalty for not satisfying the contaminant removal objective, and γ is a function of the concentration field, c, and a threshold (action) level, $\bar{c}$, which quantifies the failure to achieve the remediation objective.

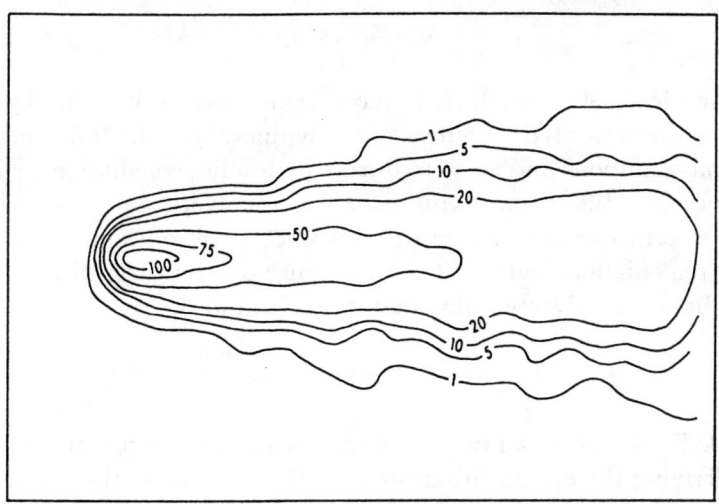

Figure 3: Hypothetical contaminant plume subjected to optimization for remedial pumping design.

In this example, we take γ to be

$$\gamma = \begin{cases} 0 & c_p \leq \bar{c} \\ c_p - \bar{c} & c_p > \bar{c} \end{cases} \tag{6}$$

where $c_p = \max(c)$ over the domain Ω at the end of the remediation period. We choose $\bar{c} = 25$ ppb (about an 80% reduction of the peak concentration) and require that remediation occur within three years. Ten (10) potential well sites distributed throughout the spatial domain are used, each having the seven potential pumping rates {-0.1, -0.05, 0.0, 0.1, 0.25, 0.5, 1.0} millions of gallons per day (mgd) (two injection, no pumping, and four extraction). The pumping rates are fixed throughout the remediation period. The resulting combinatorial problem has $7^{10} \approx 3 \times 10^8$ possible configurations. The cost coefficients are $C_{Q_i^+} = 1.0$, $C_{Q_i^-} = 10.0$, $C_{I_i} = 10.0$, and $P = 100.0$.

The evaluation of the cost function requires the evaluation of Q on the concentration field c. The governing equations for the two-dimensional model of this example are, for flow

$$S_s b \frac{\partial h}{\partial t} - Kb\nabla^2 h = Q \tag{7}$$

and, for nonreactive solute transport,

$$\frac{\partial c}{\partial t} = \boldsymbol{\nabla} \cdot (\mathbf{D} \cdot \boldsymbol{\nabla} c) - \boldsymbol{\nabla} \cdot (\mathbf{v}c) + \Gamma, \tag{8}$$

where Darcy's law gives

$$\mathbf{v} = -\frac{\mathsf{K}}{\phi}\nabla h. \tag{9}$$

Here h is the piezometric head, S_s is the specific storage, K is the hydraulic conductivity, b is the saturated thickness of the aquifer, Q is the fluid sink term (zero under natural conditions and nonzero during hydraulic remediation), c is the solute concentration, $\mathbf{D}$ is the dispersion tensor, $\mathbf{v}$ is the average pore velocity, Γ is the solute source term (nonzero under the initial 30-year contamination period, and zero during remediation), and ϕ is the uniform porosity. The dispersion tensor is assumed to be of the advection-dominated, Fickian form

$$\mathsf{D}_{ij} = \alpha_T v \delta_{ij} + (\alpha_L - \alpha_T)\frac{v_i v_j}{v}, \tag{10}$$

with $v = |\mathbf{v}|$. Because $\mathbf{D}$ is a function of Q, c is a nonlinear function of Q; the simulation underlying the optimization problem is nonlinear in the decision variables Q. Thus, a numerical simulation of flow and solute transport must be performed to evaluate γ in the contamination problem for each evaluation of the cost function. This problem therefore requires much greater computing resources.

The flow problem is solved by a node-centered finite difference method on a 25×17 uniform mesh. Solute transport is solved using a random walk particle method [Tompson and Dougherty, 1988] which includes the effects of counter-dispersion, a term corresponding to $\nabla \cdot \mathbf{D}$ in the deterministic step of the particle motion, and otherwise is essentially the same as the well-known code of Prickett *et al.* [1981].

The Monte Carlo experiments for this example comprise up to 100 replicates of simulated annealing optimization at each of the Markov chain lengths 5, 10, 20, 35, 50, 70, and 100. Because of the need for simulation for each cost evaluation, these Monte Carlo experiments required large amounts of computing time, on the order of 200 cpu hours on a Convex C240. In some cases 100 replicates were not obtained, but the missing runs did not affect our results.

Figure 4 present the results in the same format as Figures 1 and 2. We note that the average optimized cost becomes stable with chain lengths of greater than about 50. Further the percent of individual runs that obtain the minimum cost ever observed in our experiments becomes constant (about 1 in 6) for chain lengths greater than 50. Figure 4 also indicates the percentage of results within 20% of the minimum cost; this again stabilizes for chain lengths greater than 50 at about 50% (that is, every other run will be within 20 percent of the optimal solution). Finally, we note that almost every run is within 40% of the minimum cost for chain lengths of 50 or more.

Therefore, chain lengths of greater than 50 will provide consistently good results for this example. This corresponds to only 5 times the number of decision variables (here, wells). In our earlier work on this example problem [Dougherty and Marryott, 1990] we used 70 times the number of wells. We conclude that chain lengths can be about an order of magnitude shorter than previously considered while achieving good results. Because simulation costs dominate the optimization effort, this result implies an order of magnitude reduction in the amount of computational effort required during the optimization process. We do not claim this will apply in all

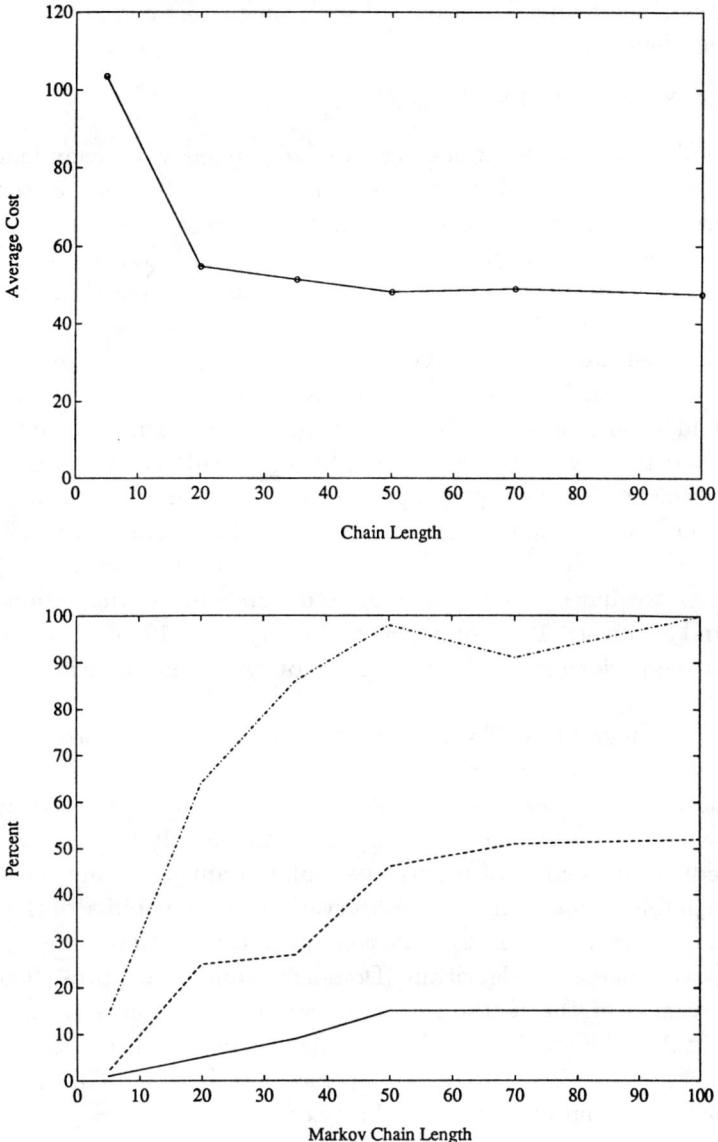

Figure 4: Results of Monte Carlo experiments on contamination example. (a) average costs of optimized solution versus Markov chain length and (b) percent of solutions attaining exactly (solid), within 20% (dash) , and within 40% (dash-dot) of best observed solution versus Markov chain length.

groundwater remediation problems. It suggests, however, that simulated annealing optimization can be effectively employed with much less computational resource than previously believed.

4. SUMMARY AND CONCLUSIONS

The method of simulated annealing can be applied with confidence to optimization problems in groundwater management. The cost function that must be evaluated during the annealing process is quite expensive to calculate, in general, because of the need to perform an extensive simulation. Biasing and other techniques may be introduced into the simulated annealing algorithm to attempt to reduce the number of simulations necessary [Dougherty and Marryott, 1990]. Another method, which we have explored here, is to seek experience on the minimum necessary length of the Markov chains within the simulated annealing algorithm which will yield good results. This result cannot be determined analytically and may be expected to vary with the problem being considered, so numerical experiments were performed for the type of applications common in subsurface hydrology.

Guidance on Markov chain length appearing in the literature (e.g., Kirkpatrick *et al.* [1983], Kirkpatrick [1984], Huang *et al.* [1986]) is rather vague. Using these guidelines we have performed simulated annealing optimization effectively [Dougherty and Marryott, 1990; Marryott and Dougherty, 1990]. The current study has examined the performance of shorter Markov chain lengths using Monte Carlo experiments.

The examples suggest that Markov chain lengths can be a full order of magnitude shorter for problems of interest to us than suggested in the literature and by our previous approach. Because the cost of optimization by simulated annealing is dominated (over 90%) by the cost of simulations, this result will lead to an order of magnitude reduction in costs of practical simulated annealing optimization. This result, when combined with our earlier observations on the ability of the method to find solutions with stringent limits on the constraints and with biasing certain stages of the simulated annealing algorithm [Dougherty and Marryott, 1990], increase the competitiveness of the method *vis-a-vis* gradient methods (e.g., Gorelick *et al.* [1984] and Ahlfeld *et al.* [1988]) and augurs well for practical optimization of groundwater and environmental management problems. This is particularly important for the evaluation of the reliability of remediation designs, in which the uncertainty of aquifer material coefficients leads to a stochastic solute transport equation.

5. ACKNOWLEDGEMENTS

We are grateful to UCI's Office of Academic Computing and its Advanced Scientific Computing group for providing us with time on its Convex C240 computer, without which our numerical experiments would be still running.

6. REFERENCES

1. E. H. L.Aarts and P. J. M. van Laarhoven (1985), "Statistical Cooling: A General Approach to Combinatorial Optimization Problems", *Phillips J. of Res.*, **40**, pp. 193–226.

2. D. P. Ahlfeld, J. M. Mulvey, G. F. Pinder, and E. F. Wood (1988), "Contaminated Groundwater Remediation Design Using Simulation, Optimization, and Sensitivity Theory, 1, Model Development", *Water Resources Research*, **24**, pp. 431–441.

3. D. E. Dougherty and R. A. Marryott (1990), "Optimal Groundwater Management, 1: Simulated Annealing", in review.

4. S. M. Gorelick, C. I. Voss, P. E. Gill, W. Murray, M. A. Saunders, and M. H. Wright (1984), "Aquifer Reclamation Design: The Use of Contaminant Transport Simulation Combined with Nonlinear Programming", *Water Resources Research*, **20**, pp. 415–427.

5. M. D. Huang, F. Romeo and A. Sangiovanni-Vincentelli (1986), "An Efficient General Cooling Schedule for Simulated Annealing", *IEEE Transactions on Computer-Aided Design*, CAD-5, pp. 381–384.

6. S. Kirkpatrick (1984), "Optimization by Simulated Annealing: Quantitative Studies", *J. of Stat. Phys.*, **34**, pp. 975–986.

7. S. Kirkpatrick, C. Gelatt, and M. Vecchi (1983), "Optimization by Simulated Annealing", *Science*, **220**:4598:671–680.

8. S. Lin (1965), "Computer Solutions of the Traveling Salesman Problem", *Bell Sys. Tech. J.*, **44**, pp. 2245–2269.

9. R. A. Marryott and D. E. Dougherty (1990), "Optimal Groundwater Management: 2. Application to Field-Scale Contamination Site", in preparation.

10. D. Mitra, F. Romeo, and A. Sangiovanni-Vincentelli (1985), "Convergence and Finite-Time Behavior of Simulated Annealing", IEEE Proc. 24th Conf. Decision and Control, Ft. Lauderdale, December, pp. 761–767.

11. T. A. Prickett, T. G. Naymik, and C. G. Lonnquist (1981), "A Random-Walk Solute Transport Model for Selected Groundwater Quality Evaluations", Illinois State Water Survey, Bulletin 65, Champaign, Ill., 103 p.

12. A. F. B. Tompson and D. E. Dougherty (1988), "On the Use of Particle Tracking Methods for Solute Transport in Porous Media", *Computational Methods in Water Resources, Volume 2: Numerical Methods for Transport and Hydrologic Processes*, M. A. Celia *et al* (eds.), Computational Mechanics Publications and Elsevier, pp. 227–232.

CHAPTER 6

Localized Adjoint Methods: A New Discretization Methodology

Ismael Herrera*

ABSTRACT. Localized Adjoint Method is a new and promising methodology of wide applicability, based on Herrera's Algebraic Theory of Boundary Value Problems. Thus, the general theory is briefly explained and then its application is illustrated with transport diffusion problems for which the Eulerian-Lagrangian Localized Adjoint Method (ELLAM) has been formulated by the LAM group (M.A.Celia, R.E. Ewing and T.F. Russell, in addition to the author). The ELLAM development unifies characteristic methods, treats boundary conditions systematically, yielding conservative schemes.

1. INTRODUCTION. The Localized Adjoint Method (LAM) is a new and promising methodology for discretizing partial differential equations which has been introduced by Herrera [1-7] and coworkers, and which is being applied to a wide range of problems [8-19]. The basic algebraic theory was developed by Herrera et. al. in [1-5] and then appplied to develop highly accurate algorithms for ordinary differential equations in [4, 8, 9]. Celia et. al. [12] developed efficient algorthms for multi-dimensional steady state problems as well as one-dimensional advection-dispersion problems [10, 11]. More recently, the LAM group (M.A. Celia, R.E. Ewing, I. Herrera and T.F. Russell) developed the Eulerian-Langrangian Localized Adjoint Method (ELLAM) [6, 7, 17, 18]. Multiphase flow simulation has been dealt with by Herrera and Ewing [13] and Ewing and Celia [14]. Contaminant transport has been considered by Celia et. al. [11, 16, 17]. Multi-dimensional nested grids are being developed by Neuman [19]. In the present article the general methodology is briefly explained, introducing new explicit formulas for the case when the coefficients of the differential operators are continuous. The application of the procedures are illustrated with the transport diffusion equations [6, 7].

The numerical solution of the advective-diffusive transport equation

---

*Instituto de Geofísica, UNAM. Mailing Address: Apdo. Postal 22-582, 14000 México, D.F., México

is a problem of great importance because many problems in science and engineering involve such mathematical model. The numerical treatment of derive from two main approaches: standard semidiscretization and Eulerian-Lagrangian. The main distinguishing feature of the latter is the use of characteristics to carry out the discretization in time. Most formulas that have been developed using a standard semidiscretization approach have been based on up-stream weighting techniques, whose development is essentially ad-hoc. This is in contrast with LAM approach which is very systematic.

There is a point on terminology, that must be mentioned. In some past work "Localized Adjoint Methods" were called "Optimal Test Function Methods", and it has been only more recently that the new terminology has been used, since it is more precise and also, it more clearly distinguishes this method from other procedures.

The starting point of Localized Adjoint Methods is a very simple but fundamental question. What is the relation between an approximate solution and the exact one. To be more precise, we proceed to give an answer to this question in a very simple situation. Consider the problem of solving the equation

$$\mathcal{L}u = f_{\Omega} , \qquad \text{in} \quad \Omega \qquad (1.1)$$

subjected to homogeneous boundary conditions for which Green's formula

$$\int_{\Omega} v\mathcal{L}u\,dx = \int_{\Omega} u\mathcal{L}^{*} v\,dx \qquad (1.2)$$

applies, when $\mathcal{L}^{*}$ is the formal adjoint of $\mathcal{L}$.. In the method of weighted residuals, one usually considers a system of weighting (or test) functions $\{w_1,...,w_N\}$. Then, a function u' is said to be an approximate solution of this problem when

$$\int_{\Omega}(\mathcal{L}u' - f_{\Omega})w_{\alpha}dx = 0 , \qquad \alpha = 1,...,N. \qquad (1.3)$$

Generally, the system of N equations (1.3) has many solutions, but in order to obtain a system possessing a unique solution, it is customary to introduce a representation $u' = \Sigma A_{\alpha}\phi_{\alpha}$ of the approximate solution in terms of the system $\{\phi_1,...,\phi_N\}$ of base (or trial) functions. However, this representation is an artifice that bears little relation with the <u>exact solution</u> u.

The following observations permit establishing the actual relation that exists between an approximate solution and the exact one and derive the actual information about the exact solution which is contained in an approximate one. From (1.1), it is clear that the exact solution u, satisfies

$$\int_{\Omega}(\mathcal{L}u - f_{\Omega})w_{\alpha}dx = 0, \qquad \alpha = 1,...,N. \qquad (1.4)$$

Equations (1.3) and (1.4) together imply

$$\int_{\Omega} w_{\alpha}\mathcal{L}u'\,dx = \int_{\Omega} w_{\alpha}\mathcal{L}u\,dx \qquad \alpha = 1,...,N. \qquad (1.5)$$

or

$$\int_{\Omega} u'\mathcal{L}^{*} w_{\alpha}\,dx = \int_{\Omega} u\mathcal{L}^{*} w_{\alpha}\,dx \qquad \alpha = 1,...,N. \qquad (1.6)$$

by virtue of Green's formula (1.1). Consider the Hilbert space L^2, of square integrable functions and in which the inner product of two functions, u and v, is given by $\int_{\Omega} uv\,dx$. Then, the system of equations (1.6) allows the following interpretation:

A function u' is an approximate solution if and only if, its projection on the space spanned by the system of functions

$\{\mathcal{L}^* w_1, ..., \mathcal{L}^* w_N\}$, *coincides with that of the exact solution u.*

As a matter of fact, this is all the information *about the exact solution contained in an approximate one.*

In this light, the representation $u' = \Sigma A_\alpha \phi_\alpha$ can be interpreted as a procedure for extrapolating the actual information contained in the approximate solution.

The very simple and precise result just presented clarifies much the nature of approximate solutions and it would be desirable to apply it, in a systematic manner, to analyze discrete methods. For this purpose it is necessary to have available Green's formulas similar to (1.2), but that can be applied even when the functions considered are not smooth, since in most numerical applications the weighting functions are localized (i.e., they have local support) and they usually do not satisfy the smoothness requirements at the boundary of their support. Even more, the development of a theory applicable to carry out the analysis when both base and test functions are fully discontinuous, is most desirable since standard theory of distributions is not applicable to that case.

Herrera [1-5], recently developed an "algebraic theory of boundary value problems" with precisely that property; that is, in which the analysis can be carried out when both trial and test functions are fully discontinuous. Such setting is ideal for localizing the adjoint equation (1.6).

"Localized adjoint methods (LAM)", in whose development the LAM group (M.A Celia, R.E. Ewing, I. Herrera and T.F. Russell) has been working and it is being applied at present to many problems [8-19], consist in making systematic use of that theory to analyze the information contained in approximate solutions. Since the quality of the results obtained with a numerical method depends, in an important manner, on the weighting functions used, one of the main goals of localized adjoint methods, thus far, has consisted in developing improved weighting functions. In this paper the LAM methodology is explained in connection with transport diffusion problems.

2. GREEN-HERRERA FORMULAS. The main ingredients of the author's Algebraic Theory of (initial) boundary value problems are

i).- General Green-Herrera formulas for differential operators in discontinuous fields; and
ii).- The operator extension (purely algebraic) induced by such formulas.

For simplicity, attention will be restricted to the case when the differential operators possess continuous coefficients, but such formulas have been developed for the general case of differential operators with discontinuous coefficients [2,4,20]. By definition, a differential operator and its formal adjoint $\mathcal{L}^*$, satisfy

$$v\mathcal{L}u - u\mathcal{L}^* v = \nabla \cdot \{\underline{\mathcal{D}}(u,v)\} \qquad (2.1)$$

where $\underline{\mathcal{D}}(u,v)$ is a vector-valued bilinear function, defined at every point of the region Ω. When considering time-dependent problems, Ω will be a region in space-time. Generally, Ω will be divided in many subregions (the elements), in each of which weighting and test functions will be assumed to be sufficiently differentiable for the operators to be well defined in its interior. The union of the interelement boundaries will be denoted by Σ. Trial and test functions will be taken from two linear spaces D_1 and D_2, and they together with their derivatives may have jump discontinuities across Σ.

Integration of equation (2.1) and application of generalized divergence (Gauss) theorem [21], yields:

$$\int_\Omega \{v\mathcal{L}u - u\overset{*}{\mathcal{L}}v\}dx = \int_{\partial\Omega^-} \underline{\mathcal{D}}(u,v)\cdot\underline{n}\ dx - \int_\Sigma [\underline{\mathcal{D}}(u,v)]\cdot\underline{n}\ dx \qquad (2.2)$$

Generally, $\mathcal{L}u$ as well as $\overset{*}{\mathcal{L}}v$ may not be defined on Σ, where the functions may be discontinuous. Thus, here as in what follows, integrals over Ω are carried out excluding Σ and differential operators are understood in an elementary sense and not in a distributional sense. In equation (2.2), the square brackets stand for the "jumps" across Σ of the function contained inside; i.e. value on the plus side minus value in the minus side (the plus side is defined as that one towards which the unit normal on Σ points to).

Green–Herrera formulas are obtained by carrying out suitable decompositions of the bilinear functions $\underline{\mathcal{D}}(u,v)\cdot\underline{n}$ on the boundary $\partial\Omega$ of Ω and of $-[\underline{\mathcal{D}}(u,v)]\cdot\underline{n}$ on the interelemnt boundaries Σ. The decomposition of $\underline{\mathcal{D}}(u,v)\cdot\underline{n}$ is standard (see, for example, Lions and Magenes [22]) and leads to the definition of two bilinear functions $\mathcal{B}(u,v)$ and $\mathcal{C}(v,u)$ such that

$$\underline{\mathcal{D}}(u,v)\cdot\underline{n} = \mathcal{B}(u,v) - \mathcal{C}(v,u) \qquad (2.3)$$

whose definitions depend on the type of boundary and initial conditions, which are prescribed. The function $\mathcal{B}(u,v)$ is such that, when considering boundary (initial) value problems, for any u which satisfies the prescribed boundary conditions, $\mathcal{B}(u,v)$ is a well-defined linear function of v, independent of the particular choice of u. This linear function will be denoted by g_∂ (thus, its value for any given function v, will be $g_\partial(v)$).

The decomposition of $-[\underline{\mathcal{D}}(u,v)]\cdot\underline{n}$ is easily carried out using the identity

$$[\underline{\mathcal{D}}(u,v)] = \underline{\mathcal{D}}_+(u,v) - \underline{\mathcal{D}}_-(u,v) = \underline{\mathcal{D}}(u_+,v_+) - \underline{\mathcal{D}}(u_-,v_-) \qquad (2.4)$$

which holds when the differential operators have continuous coefficients. Let the average $\overset{\bullet}{u}$ of any function u be defined by

$$\overset{\bullet}{u} = (u_+ + u_-)/2 \qquad (2.5)$$

Then it is easy to verify that

$$u_+ = \overset{\bullet}{u} + \tfrac{1}{2}[u] ; \qquad u_- = \overset{\bullet}{u} - \tfrac{1}{2}[u] \qquad (2.6)$$

which allows writing

$$\underline{\mathcal{D}}(u_+,v_+) = \underline{\mathcal{D}}(\overset{\bullet}{u},\overset{\bullet}{v}) + \tfrac{1}{2}\underline{\mathcal{D}}(\overset{\bullet}{u},[v]) + \tfrac{1}{2}\underline{\mathcal{D}}([u],\overset{\bullet}{v}) + \tfrac{1}{4}\underline{\mathcal{D}}([u],[v]) \qquad (2.7a)$$

$$\underline{\mathcal{D}}(u_-,v_-) = \underline{\mathcal{D}}(\overset{\bullet}{u},\overset{\bullet}{v}) - \tfrac{1}{2}\underline{\mathcal{D}}(\overset{\bullet}{u},[v]) - \tfrac{1}{2}\underline{\mathcal{D}}([u],\overset{\bullet}{v}) + \tfrac{1}{4}\underline{\mathcal{D}}([u],[v]) \qquad (2.7b)$$

Equations (2.4) and (2.7) together, yield

$$[\underline{\mathcal{D}}(u,v)] = \underline{\mathcal{D}}(\overset{\bullet}{u},[v]) + \underline{\mathcal{D}}([u],\overset{\bullet}{v}) \qquad (2.8)$$

Defining

$$\mathcal{J}(u,v) \equiv -\underline{\mathcal{D}}([u],\overset{\bullet}{v})\cdot\underline{n} \qquad \mathcal{K}(v,u) \equiv \underline{\mathcal{D}}(\overset{\bullet}{u},[v])\cdot\underline{n} \qquad (2.9)$$

it is clear that

$$-[\underline{D}(u,v)]\cdot\underline{n} = \mathcal{J}(u,v) - \mathcal{K}(v,u) \qquad (2.10)$$

This is the desired decomposition of the bilinear function $-[\underline{D}(u,v)]\cdot\underline{n}$. The basic properties which lead to this choice of $\mathcal{J}$ and $\mathcal{K}$, have been developed more thoroughly in the author's algebraic theory [2]. In particular, at a given point of Σ, $[u]$ and $\dot{u}$, can be varied independently. When $[u]$ takes a given value, $\mathcal{J}(u,v)$ is a well defined functional of v, independent of the specific choice of u. In connection with boundary (initial) value problems with prescribed jumps, the linear function of v which is obtained when $\mathcal{J}(u,v)$ is evaluated keeping $[u]$ equal to the prescribed jumps, will be denoted by j_∂ (thus, its value for any given v, will be $j_\partial(v)$).

At this point it is convenient to introduce the following bilinear functionals:

$$\langle Pu,v\rangle \equiv \int_\Omega v\mathcal{L}u\,dx\,; \quad \langle Qv,u\rangle \equiv \langle \overset{*}{Q}u,v\rangle \equiv \int_\Omega u\mathcal{L}^* v\,dx \qquad (2.11a)$$

$$\langle Bu,v\rangle \equiv \int_{\partial\Omega}\mathcal{B}(u,v)\,dx; \quad \langle \overset{*}{C}u,v\rangle \equiv \int_{\partial\Omega}\mathcal{C}(u,v)\,dx \qquad (2.11b)$$

$$\langle Ju,v\rangle \equiv \int_\Sigma \mathcal{J}(u,v)\,dx; \quad \text{and} \quad \langle \overset{*}{K}u,v\rangle \equiv \int_\Sigma \mathcal{K}(v,u)\,dx \qquad (2.11c)$$

By means of these definitions, using (2.3) and (2.10), equation (2.2) becomes

$$P - \overset{*}{Q} = B - \overset{*}{C} + J - \overset{*}{K} \qquad (2.12)$$

This is <u>Green–Herrera</u> <u>formula</u> <u>for</u> <u>operators</u> <u>in</u> <u>discontinuous</u> <u>fields</u> [2,5]. It can be applied when both trial and test functions are discontinuous, something which is not possible when using standard theory of distributions.

Rearranging (2.12), it is possible to write

$$P - B - J = \overset{*}{Q} - \overset{*}{C} - \overset{*}{K} \qquad (2.13)$$

The left–hand side of equation (2.13) can be interpreted as an extension of the differential operator $\mathcal{L}$, which was originally defined for differentiable functions only, to fully discontinuous functions. Indeed:

> *The differential operator $\mathcal{L}$, when applied to a*
> *smooth function u which stisfies homogeneous*
> *boundary conditions, yields the linear functional*
> Pu. *However, when applied to a fully discontinuous*
> *function u, which does not satisfy homogeneous*
> *boundary conditions, it yields the linear functional*
> (P–B–J)u.

The general boundary (initial) value problem to be considered is one with prescribed jumps. The differential equation is

$$\mathcal{L}u = f_\Omega \qquad \text{in } \Omega \qquad (2.14)$$

In addition, some boundary conditions (and initial conditions) are specified on $\partial\Omega$, while jump conditions are specified at Σ. When modelling continuous systems, such jump conditions stem from basic conservation or, more generally, balance laws of continuous mechanics [20].

Using the bilinear functionals thus far introduced, such boundary (initial) value problem with prescribed jumps, can be formulated variationally as

$$\langle Pu,v\rangle = \langle f,v\rangle; \quad \langle Bu,v\rangle = \langle g,v\rangle; \quad \langle Ju,v\rangle = \langle j,v\rangle \;\forall\; v\epsilon D_2 \qquad (2.15)$$

where f, g and j, are linear functionals defined by

$$\langle f,v \rangle = \int_\Omega v f_\Omega dx \; ; \; \langle g,v \rangle = \int_{\partial\Omega} g_\partial(v) dx \; ; \; \langle j,v \rangle = \int_\Sigma j_\partial(v) \, dx \quad (2.16)$$

All they can evaluated using the data of the problem.

Generally, the bilinear functionals J and B, are <u>boundary</u> <u>operators</u> for P, which are <u>fully</u> <u>disjoint</u> (For the definitions of the concepts that have been underlined here, the reader is referred to the author's original papers [2,5]. Further details can also be found in those publications). When this is the case, the system of equations (2.15), is equivalent to the single variational equation

$$\langle (P - B - J)u,v \rangle = \langle f - g - j,v \rangle \qquad \forall \; v \varepsilon D_2 \quad (2.17)$$

This is said to be "the variational formulation in terms of the data of the problem", becuase Pu, Bu and Ju are prescribed. Making use of Green–Herrera formula (2.13), the variational formulation (2.17) is transformed into

$$\langle (Q^* - C^* - K^*)u,v \rangle = \langle f - g - j,v \rangle \; \forall \; v \varepsilon D_2 \quad (2.18)$$

This is said to be "the variational formulation in terms of the sought information", because $Q^* u$, $C^* u$ and $K^* u$ are not prescribed. It can be seen, by virtue of equations (2.11), that $Q^* u$, $C^* u$ and $K^* u$ supply information about the sought solution at the interior of the region Ω (where the problem is defined), the complementary boundary values at $\partial\Omega$ and the average of the solution (and its derivatives) across the surfaces Σ of discontinuity.

Making use of the variational formulation in terms of the sought information, the arguments that lead to the formulation of Localized Adjoint Methods, constitute a mere repetition of those presented in the Introduction. Given a system of weighting functions $\{w_1,...,w_N\} \subset D_2$, an approximate solution is again any function $u' \varepsilon D_1$ which satisfies

$$\langle (Q^* - C^* - K^*)u',w_\alpha \rangle = \langle f-g-j,w_\alpha \rangle \; , \; \alpha=1,...,N \quad (2.19)$$

Clearly, this equation together with (2.18) implies

$$\langle (Q^* - C^* - K^*)u',w_\alpha \rangle = \langle (Q^* - C^* - K^*)u,w_\alpha \rangle, \; \alpha=1,...,N \quad (2.20)$$

since an exact solution also satisfies (2.19). Equation (2.20), is the basis for the analysis of the information contained in an approximate solution and constitutes the starting point of Localized Adjoint Methods.

3. EULERIAN-LAGRANGIAN LAM. When applying the methods of Section 2 to time dependent problems, it will be necessary to consider a region Ω in space-time. Also, the surface Σ on which discontinuities can occur, will be a surface in space-time and a suitable notation will be required. The set of points of Ω whose time value is t, will be denoted by $\Omega(t)$ and correspondingly, $\Sigma(t)$ stands for the set of points of Σ whose time value is t. Space-time vectors $\underline{M}$ will be written as pairs:

$$\underline{M} = (\underline{m} , m_t) \qquad (3.1)$$

where $\underline{m}$ is the vector made by its spatial components and while m_t corresponds to its temporal component. Let the $\underline{V}_\Sigma$ be the vectorial velocity of the surface of discontinuity $\Sigma(t)$. This is a space vector which can be written as

$$\underline{V}_\Sigma = V_\Sigma \underline{n} \qquad (3.2)$$

where $\underline{n}$ is the unit normal vector to $\Sigma(t)$. Generally, V_Σ can be positive or negative, depending on the sense of motion of $\Sigma(t)$ and the choice of $\underline{n}$.

3.1.– **One Dimensional ELLAM.**– For this case, $\underline{n}$ can be chosen as equal to one and this will be done in what follows. Observe that the space–time vector $(V_\Sigma,1)$ is tangent to Σ. Using this fact, it is easy to see that a space–time unit normal vector $\underline{N}$ to Σ, is given by

$$\underline{N} = (1+V_\Sigma^2)^{-1/2}(1, -V_\Sigma) \qquad (3.3)$$

Consider the one-dimensional transient advection–diffusion equation subject to appropiate initial and boundary conditions

$$\mathcal{L}u \equiv \frac{\partial u}{\partial t} + V\frac{\partial u}{\partial x} - D\frac{\partial^2 u}{\partial x^2} = f_\Omega(x,t), \text{ in } \Omega \qquad (3.4)$$

$$u(x,0) = u_I(x), \qquad \text{on } \partial_I\Omega \equiv \Omega(0) \qquad (3.5a)$$

$$u(0,t) = u_o(t), \qquad \text{on } \partial_o\Omega \qquad (3.5b)$$

$$\frac{\partial u}{\partial x}(l,t) = q_l(t), \qquad \text{on } \partial_l\Omega \qquad (3.5c)$$

Here, the space–time region $\Omega = \Omega_x \times \Omega_t$, with $\Omega_x = [0,l]$ and $\Omega_t = [0,T]$. In addition, $\partial_o\Omega$ and $\partial_l\Omega$ are the subsets of Ω for $x=0$ and $x=l$, respectively. First and second type boundary conditions are assumed for demonstration purposes only; the following development accomodates any combination of boundary conditions.

The adjoint operator is

$$\mathcal{L}^*w \equiv -\frac{\partial w}{\partial t} - \frac{\partial}{\partial x}(Vw) - D\frac{\partial^2 w}{\partial x^2} \qquad (3.6)$$

and $\mathcal{D}(u,v)$ as defined by (2.5) is

$$\mathcal{D}(u,w) \equiv \left\{ uD\frac{\partial w}{\partial x} - w\left(D\frac{\partial u}{\partial x} - Vu\right), uw \right\} \qquad (3.7)$$

Therefore

$$-[\mathcal{D}(u,w)]\cdot\underline{N} = -(1+V_\Sigma^2)^{-1/2}\left[uD\frac{\partial w}{\partial x} - w\left(D\frac{\partial u}{\partial x} - (V-V_\Sigma)u\right)\right] \qquad (3.8)$$

Assuming that the physical process which equation (3.4) mimics is that of transport with Fickian diffusion, of a solute whose concentration is u, in a free fluid moving with velocity V, the smoothness conditions implied by mass balance are

$$[u(V - V_\Sigma) - D\frac{\partial u}{\partial x}] = 0, \qquad \text{on } \Sigma \qquad (3.9a)$$

In addition, Fickian diffusion implies

$$[u] = 0, \qquad \text{on } \Sigma \qquad (3.9b)$$

When the coefficients V and D are continuous, equations (3.9) are satisfied, if and only if, u and its spatial derivative are continuous across Σ. Application of equation (2.9) yields

$$\mathcal{J}(u,w) = -(1+V_\Sigma^2)^{-1/2}\left\{[u]D\overline{\frac{\partial w}{\partial x}} - \dot{w}\left(D\left[\frac{\partial u}{\partial x}\right] - (V-V_\Sigma)[u]\right)\right\} \qquad (3.11a)$$

$$\mathcal{K}\,(w,u) = (1+V_\Sigma^2)^{-1/2}\left\{\dot{u}\left[D\frac{\partial w}{\partial x}\right] - [w]\left(D\overline{\frac{\partial u}{\partial x}} - (V - V_\Sigma)\dot{u}\right)\right\} \qquad (3.11b)$$

It is clear now that $j_\Sigma(v) \equiv 0$ on Σ, since the sought solution is required to satisfy the condition $[u] = \left[\frac{\partial u}{\partial x}\right] \equiv 0$ on Σ.

The bilinear functions $\mathcal{C}(u,w)$ and $\mathcal{B}(u,w)$, to be defined on the boundary, must be constructed taking into account the kind of boundary conditions to be satisfied. For the kind of boundary conditions given by equations (3.5), they are

$$\mathcal{B}(u,w) = u\left(D\frac{\partial w}{\partial x} + V\,w\right), \quad \mathcal{C}(w,u) = wD\frac{\partial u}{\partial x}, \quad \text{on } \partial_o\Omega \qquad (3.12a)$$

$$\mathcal{B}(u,w) = -\,wD\frac{\partial u}{\partial x}, \quad \mathcal{C}(w,u) = -u\left(D\frac{\partial w}{\partial x} + V\,w\right), \text{ on } \partial_l\Omega \qquad (3.12b)$$

$$\mathcal{B}(u,w) = -uw \qquad \text{on } \partial_I\Omega \qquad (3.12c)$$

$$\mathcal{C}(w,u) = -uw \qquad \text{on } \partial_T\Omega \qquad (3.12d)$$

where $\partial_T\Omega \equiv \Omega(T)$. Observe that the decomposition of $\underline{\mathcal{D}}(u,v)\cdot\underline{N}$ in $\Omega(0)\cup\Omega(T)$ does not have a point-wise character. This reflects the fact that this is an initial-value problem. From (3.5) and (3.12), it follows that

$$g_\partial(w) = u_o\left(D\frac{\partial w}{\partial x} + V\,w\right) \quad \text{on } \partial_o\Omega \qquad (3.13a)$$

$$g_\partial(w) = -\,wDq_l \qquad \text{on } \partial_l\Omega \qquad (3.13b)$$

$$g_\partial(w) = -u_I w \qquad \text{on } \partial_I\Omega \qquad (3.13c)$$

The expressions for the bilinear functionals B, C, J and K, are obtained integrating $\mathcal{B}$, $\mathcal{C}$, $\mathcal{J}$ and $\mathcal{K}$, on the boundary and on Σ. Asuming that Σ is the union of the set of curves $\{\Sigma_1, \Sigma_2,..., \Sigma_E\}$ in space-time (Fig.1),

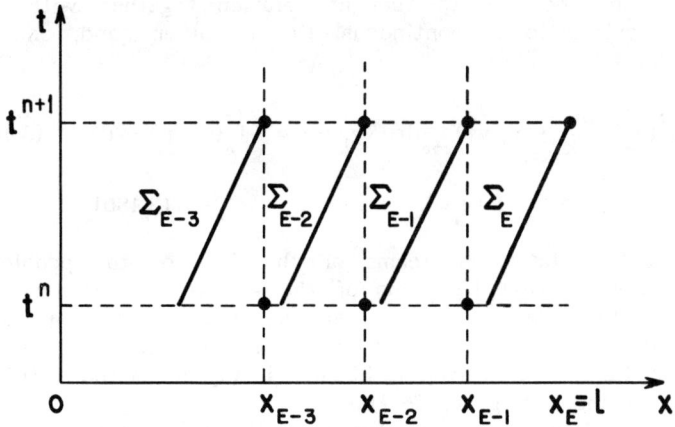

Figure 1.- The surface of discontinuity Σ.

the expressions for J and K^* are the summ of the contributions of each one of these curves. Thus, one can write

$$J = \sum_{\alpha=1}^{E} J_\alpha \qquad \text{and} \qquad K^* = \sum_{\alpha=1}^{E} K_\alpha^* \qquad (3.14)$$

where

$$\langle J_\alpha u, w\rangle = \int_0^T \left\{ [u]D\overline{\frac{\partial w}{\partial x}} - \dot{w}\left(D\left[\frac{\partial u}{\partial x}\right] - (V - V_\Sigma)[u]\right) \right\}_\alpha dt \qquad (3.15a)$$

$$\langle \overset{*}{K}_\alpha u, w\rangle = \int_0^T \left\{ \dot{u}\left[D\overline{\frac{\partial w}{\partial x}}\right] - [w]\left(D\frac{\partial u}{\partial x} - (V - V_\Sigma)\dot{u}\right) \right\}_\alpha dt \qquad (3.15b)$$

Here, the subindex α means that the value of the integrand is taken on Σ_α. To obtain equations (3.14), use has been made of the fact that, on each Σ_α, the element of time dt, is $(1+V_\Sigma^2)^{-1/2}$ times the length element in space-time.

In a similar fashion, it is convenient to decompose the bilinear functionals B and $\overset{*}{C}$ into the contributions which stem from $\partial_I\Omega$, $\partial_T\Omega$, $\partial_0\Omega$ and $\partial_l\Omega$. The corresponding expressions are:

$$B = B_I + B_o + B_l \qquad \text{and} \qquad \overset{*}{C} = \overset{*}{C}_T + \overset{*}{C}_o + \overset{*}{C}_l \qquad (3.16)$$

In view of (3.12), it is clear that

$$\langle B_o u, w\rangle = \int_0^T \left\{ u\left(D\frac{\partial w}{\partial x} + V\,w\right) \right\}_{x=o} dt; \quad \langle B_l u, w\rangle = -\int_0^T \left\{ wD\frac{\partial u}{\partial x} \right\}_{x=l} dt \qquad (3.17a)$$

$$\langle B_I u, w\rangle = -\int_0^l (uw)_{t=o} dx; \quad \langle \overset{*}{C}_T u, w\rangle = -\int_0^l (uw)_{t=T} dx \qquad (3.17b)$$

$$\langle \overset{*}{C}_o u, w\rangle = \int_0^T \left(wD\frac{\partial u}{\partial x}\right)_{x=o} dt; \quad \langle \overset{*}{C}_l u, w\rangle = -\int_0^T \left\{ u\left(D\frac{\partial w}{\partial x} + V\,w\right) \right\}_{x=l} dt \qquad (3.17c)$$

To complete the formulation of the problem, it remains to define the linear functionals f, g, and j. The first one is given by (2.16), while the latter one is zero, because the sought solution together with its normal derivative is required to be continuous. On the other hand, one can write $g = g_I + g_o + g_l$, with

$$\langle g_o, w\rangle = \int_0^T \left\{ u_o\left(D\frac{\partial w}{\partial x} + V\,w\right) \right\}_{x=o} dt; \quad \langle g_l, w\rangle = -\int_0^T \{wDq_l\}_{x=l} dt \qquad (3.18a)$$

$$\langle g_I, w\rangle = -\int_0^l (u_I w)_{t=o} dx; \qquad (3.18b)$$

The variational formulation in terms of the data of the problem is now given by (2.17), while that in terms of the sought information is given by (2.18). This equation was applied to analyze the information contained in approximate solutions. In this manner a generalization of Characteristic Methods, called Eulerian- Lagragian Localized Adjoint Methods (ELLAM), has been developed by the LAM group [6,7].

Many numerical methods use characteristic analysis to accomodate the advective component of transport. Such Characteristic Methods include Eulerian–Lagrangian Methods (ELM) [9–11], Modified Method of Characteristics (MMOC) [12,13], and operator splitting methods [14,15]. The ELLAM approximations provide a systematic framework for development of Characteristic Methods for numerical approximation of advective–diffusive transport equations. The Localized Adjoint Method (LAM) procedures lead naturally to the definition of special space–time test functions that produce the generalized CM approximations. The resulting set of approximating equations subsumes many of the CM approximations proposed in the literature. It therefore unifies these methods. In addition, the development inherently provides a systematic procedure for proper incorporation of all types of boundary conditions in a mass conservative manner.

REFERENCES

[1] Herrera, I.,"Boundary Methods: An Algebraic Theory", Pitman Advanced Publishing Program, London, 1984.

[2] Herrera, I.,"Unified Approach to Numerical Methods", Part I, Num. Methods for Partial Differential Equations 1(1), 25–44, 1985.

[3] Herrera, I.,"Unified Approach to Numerical Methods", Part II, Num. Methods for Partial Differential Equations 1(3), 159–186, 1985.

[4] Herrera, I., Chargoy, L. and Alduncin, G.,"Unified Approach to Numerical Methods", Part III, Num. Methods for Partial Differential Equations 1(4), 241–258, 1985.

[5] Herrera, I.,"Unifying Concepts in Applied Mathematics", In "The Merging of Disciplines: New Directions in Pure, Applied and Computational Mathematics", Eds. R.E. Ewing, K.I. Gross and C.F. Martin. Springer-Verlag, New York, 1986 (Invited Paper).

[6] Celia, M.A., T.F. Russell, I. Herrera and R.E. Ewing,"An Eulerian–Lagrangian Localized Adjoint Method for the Advection-Diffusion Equation", Advances in Water Resources, 13(4), pp 187–201, 1990.

[7] Herrera, I., R.E. Ewing, T.F. Russell and M.A. Celia, Eulerian–Lagrangian Localized Adjoint Methods: Theoretical Frame-work, Submitted to SIAM Journal of Numerical Analysis, 1991.

[8] Celia, M.A., and I. Herrera, "Solution of general ordinary differential equations by a unified theory approach", *Numerical Meth Partial Diff Eq.*, 3(2), pp. 117–129, 1987.

[9] Herrera, I., "The algebraic theory approach for ordinary differential equations: Highly accurate finite differences", *Numerical Meth Partial Diff Eq.*, 3(3), pp. 199–218, 1987.

[10] Celia, M.A., Herrera I., Bouloutas, E.T., and Kindred, J.S., "A new numerical approach for the advective-diffusive transport equation", *Numerical Meth Partial Diff Eq.*, 5, pp. 203–226, 1989.

[11] Celia, M.A., Kindred, J.S. and Herrera I., "Contaminant transport and biodegradation, 1, A numerical model for reactive transport in porous media", Water Resour. Res., 25(6), pp. 1141–1148, 1989.

[12] Celia, M.A., Herrera I., Bouloutas, E.T., "Adjoint Petrov Galerkin methods for multi-dimensional flow problems", In *Finite Element Analysis in Fluids,* T.J. Chung and R. Karr Eds., UAH Press, Hunstville, Alabama, pp. 953–958, 1989. (Invited Paper).

[13]   Herrera, I. and Ewing, R.E. "Localized Adjoint Methods:
Applications to Multiphase Flow Problems".   Proceedings Fifth
Wyoming Enhanced Oil Recovery Symposium, May 10-11, 1989,
Casper Wyoming, pp.155-173, 1990.

[14]   Ewing, R.E. and Celia. M.A., "Multiphase Flow Simulation in
Groundwater   Hydrology   and   Petroleum   Engineering".
Computational   Methods   in   Subsurface   Hydrology,   Eds,   G.
Gambolati   et   al.,   Computatinal   Mechanics   Publications,
Springer Verlag, pp. 195-202. 1990.

[15]   Herrera, I., "Localized Adjoint Methods in Water Resources
Problems" Computational Methods in Surface Hydrology, Eds.
G.   Gambolati   et   al.,   Computatinal   Mechanics   Publications,
Springer Verlag, pp. 433-440. 1990 (Invited Paper).

[16]   Zisman, S., "Simulation of contaminant transport in groundwater
systems using Eulerian-Langrangian localized adjoint methods,"
MS Thesis, Dept. Civil Eng., MIT, 1989.

[17]   Celia, M.A and Zisman S,"Eulerian-Langrangian Localized
Adjoint   Method   for   Reactive   Transport   in   Groundwater"
Computational   Methods   in   Subsurface   Hydrology,   Eds,   G.
Gambolati   et   al.,   Computatinal   Mechanics   Publications,
Springer Verlag, pp. 383-390. 1990.

[18]   Russell, T.F and R.V. Trujillo., "Eulerian-Lagrangian
Localized   Adjoint   Methods   with   Variable   Coefficients   in
Multiple   Dimensions",   Computational   Methods   in   Surface
Hydrology, Eds. G. Gambolati et al., Computatinal Mechanics
Publications, Springer Verlag, pp. 357-363. 1990.

[19]   Neuman, S.P.,"Adjoint Petrov-Galerkin Method with Optimum
Weight   and   Interpolation   Functions   Defined   on
Multi-dimensional   Nested   Grids",   Computational   Methods   in
Surface   Hydrology,   Eds   G.   Gambolati   et   al.,   Computatinal
Mechanics Publications, Springer Verlag, pp. 347-356, 1990.

[20]   Herrera, I.,"Localized Adjoint Methods: Theoretical
Foundations", *Numerical Meths. Partial Diff. Eq.*, (to be
published), 1990.

[21]   Allen, M.B., I. Herrera and G.F. Pinder,"Numerical Modeling
in Science and Egineering", John Wiley, New York, 1988.

[22]   Lions, J.L. and E. Magenes,"Non-homogeneous Boundary Value
Problems and Applications", Vol. I, Springer Verlag, 1972.

[23]   Baptista, A.M.,"Solution of advection dominated transport by
Eulerian-Lagrangian methods using the backwards methods of
characteristics",Ph.D. Thesis, Dept. of Civil Engineering,
M.I.T., 1987.

[24]   Neuman, S.P.,"An Eulerian-Lagrangian numerical scheme for the
dispersion convection equation using conjugate space-time
grids", J. Comp. Phys., 41, 270-294, 1981.

[25]   Neuman, S.P.,"Adaptive Eulerian-Lagrangian finite element
method   for   advection-dispersion",   Int.   J.   Num.   Meth.
Engrg., 20, 321-337, 1984.

[26]   Ewing, R.E., T.F. Russell and M.F. Wheeler,"Convergence
analysis   of   an   approximation   of   miscible   displacement   in

porous media by mixed finite elements and a modified method of characteristics", Comp. Meth. Appl. Mech. Engrg., 47, 73–92, 1984.

[27]   Douglas, J. Jr. and T.F. Russell,"Numerical methods for convection dominated diffusion problems based on combining the method of characteristics with finite elemnt or finite difference procedures", SIAM J. Num. Anal., 19, 871–885, 1982.

[28]   Espedal, M.S. and R.E. Ewing,"Characteristic Petrov-Galerkin subdomain methods for two phase immiscible flow", Comp. Meth. Appl. Mech. Engrg., 64, 113–135, 1987.

[29]   Dahle, H.K., M.S. Espedal, R.E. Ewing and O. Saevareid,"Characteristic adaptive sub-domain methods for reservoir flow problems", Numerical Methods for Partial Differential Equations, to appear, 1990.

CHAPTER 7

A Method of Factoring Long Z-Transform Polynomials†

J. P. Lindsey*
James W. Fox**

Abstract. *Most z–transform roots of seismic data lie very near the unit circle. Also, in rho–theta coordinates, there are regular separation patterns in the theta direction. A grid search surrounding the unit circle will typically find at least 96% of all roots; polynomial division will find the rest. The Fast Fourier Transform (FFT) of length N simultaneously evaluates a polynomial at all N'th roots of unity. The FFT is thus an economical means of performing the grid search along the unit circle. Evaluation on circles with radii differing from 1.0 can be performed by multiplying the polynomial coefficients by an appropriate exponential before taking the FFT. The method is highly vectorizable and parallelizable.*

1. Introduction. Virtually all modern exploration data (seismic, well log, etc.) is acquired, processed, and archived in sampled data form. One of the principal processes employed is digital filtering which, conveniently, is equivalent to polynomial multiplication. Inverse filtering, frequently referred to broadly as "deconvolution," is roughly equivalent to polynomial division. Thus, all of polynomial algebra is directly applicable to the study of linear filtering of sampled data.

This correspondence of polynomial algebra to linear filtering was recognized early in post–war seismic research. However, its application was severely limited because polynomial root solving methods were expensive and could not handle long polynomials such as those that arise as z–transforms of seismic data. Consequently, research in digital filtering moved away from factoring long polynomials and generally settled on least–square–error and Fourier techniques.

† Funded by GeoQuest International, Inc., now Petroleum Information Corp.
* Petroleum Information Corp.
** Formerly Consultant to GeoQuest International, Inc., now Exxon Production Research Co.

The availability of an inexpensive and accurate root solving method for long polynomials (more than 100 coefficients) permits the revival of the study of polynomial methods and z–transform applications to the problems of digital filtering of exploration data. This paper discloses such a method.

2. The Fourier Transform as a Polynomial Evaluator on the Unit Circle.

If a time series is sampled to get a set of N values, (f_0 , f_1 , f_2 , f_3 , $\dots$, f_{N-1}), the Discrete Fourier Transform is defined as,

$$F_k = \sum_{j=0}^{N-1} (e^{-2\Pi i k/N})^j f_j \qquad k = 0,\ 1,\ 2,\ 3,\ \dots\ ,N-1$$

Consider the associated z–transform polynomial,

$$P(z) = \sum_{j=0}^{N-1} f_j\, z^j$$

If P(z) is evaluated at the N'th roots of unity,

$$1,\ e^{-2\Pi i\, 1/N},\ e^{-2\Pi i\, 2/N},\ \dots\ ,e^{-2\Pi i\,(N-1)/N}$$

the result for the k'th root is,

$$P(z_k) = \sum_{j=0}^{N-1} (e^{-2\Pi i k/N})^j f_j$$

which is identical to the terms of the Fourier transform.

Thus, the Fourier transform simultaneously evaluates the polynomial P(z) at all N'th roots of unity, and it does so in an extremely efficient manner.

2.1 Evaluating the Polynomial Off the Unit Circle. If the coefficients of the polynomial P(z) are multiplied by an exponential sequence:

$$1\ ,\ r\ ,\ r^2\ ,\ r^3\ ,\ \dots\ ,\ r^{N-1}$$

we get a new polynomial,

$$Q(z) = \sum_{j=0}^{N-1} r^j f_j\, z^j$$

Note that,

$$Q(z) = P(rz)$$

where z is any complex number.

In particular, $Q(e^{i\phi}) = P(re^{i\phi})$. Therefore, if r is a positive real number, evaluating Q(z) on the unit circle is equivalent to evaluating P(z) on the circle with radius r [1].

Since the Fourier transform permits us to simultaneously evaluate any polynomial at all N'th roots of unity, we now have a quick method to evaluate at N equally spaced points on any circle centered at the origin.

2.2 Evaluating the Polynomial in the Upper Half–Plane. If the polynomial has real coefficients, the work can essentially be cut in half by using the following facts:

    a) If c is any complex root of P(z), its complex conjugate, $\bar{c}$, is also a root.

    b) If c is in the upper half–plane, $\bar{c}$ is in the lower half–plane.

Therefore, if all the roots which lie in the upper half–plane or on the real axis can be found, conjugation will immediately give all the remaining roots in the lower half–plane. Fortunately, there is a version of the FFT that does exactly that.

Most mathematical subroutine libraries contain two versions of the FFT algorithm.

One version of the FFT can transform an arbitrary sequence of complex numbers. It is equivalent to evaluating the polynomial at all N'th roots of unity. If the input array is real, it must be converted to a complex array by inserting 0.0 after each sample because all the imaginary parts are zero. Thus, the original sequence of N values must be expanded to 2N values before taking the complex FFT.

The other version, the real FFT, is optimized to handle the case where the input is real. It is not necessary to insert zeros to create imaginary parts. It is equivalent to evaluating the polynomial at all N'th roots of unity in the upper half–plane. Since it deals with N points instead of 2N points, it is almost twice as fast.

2.3 Evaluating in the Upper Half–Plane, On One Side of the Unit Circle. There are situations where the work can be further cut in half because it is only necessary to search on one side of the unit circle.

If the polynomial is known to be minimum or maximum phase, there are no roots on one side of the unit circle and the search can be adjusted accordingly.

If a real polynomial is zero–phase, as for example a truncated autocorrelation, the roots outside the unit circle can easily be computed from the roots inside the circle. In polar coordinates, if $re^{i\phi}$ is a root inside the unit circle, then $(1/r)e^{i\phi}$ is also a root, and it is both outside the circle and in the same half–plane. For future reference, we will call these "reciprocal roots." (Roots on the unit circle are their own reciprocals).

The roots of a zero–phase wavelet must occur in reciprocal pairs because, if P(z) is an N'th degree symmetric polynomial, then

$$P(z) = z^N P(1/z)$$

This becomes evident by examining a general fourth degree zero–phase polynomial:

$$P(z) = a + bz + cz^2 + bz^3 + az^4$$

Note that,

$$z^4 P(1/z) = z^4(a + b/z + c/z^2 + b/z^3 + a/z^4) = P(z)$$

Therefore, if c is a root, then 1/c is also a root. If the polynomial has real coefficients, then their conjugates must also be roots. Thus, for zero–phase real polynomials, roots off the unit circle occur in groups of four, c, 1/c, $\bar{c}$, and $1/\bar{c}$.

Reciprocal roots are symmetric about the unit circle if the radial axis is plotted logarithmically; log(r) and log(1/r) are equal but opposite in polarity (log (1) = 0).

3. Plotting Polynomial Roots: The Rectangularized Z–Plane. The closeness of root positions to the unit circle for typical seismic data presents a plotting problem. Using the conventional z–plane yields a less informative plot since the root radii are not very different from 1. What is needed is a plot format that permits independent scaling of the root radii and their z–plane angle so that small variations in radii can be more easily observed. This is accomplished by rectangularizing the upper half of the z–plane as indicated in Figures 1 and 2.

The unit circle in a conventional z–plane plot is shown in Figure 1. Two roots are indicated at positions A and B. Each root has a complex conjugate paired root since the polynomial from which they came has real coefficients.

Figure 2 shows the transformed upper half of the z–plane of Figure 1 with radius plotted vertically and angle plotted horizontally. The roots A and B are shown, but not their conjugates since only the upper half of the z–plane is shown. The radius scale can thus be independently expanded relative to the angle scale for critical viewing of roots close to the unit circle. All subsequent root plots will be in the format of Figure 2.

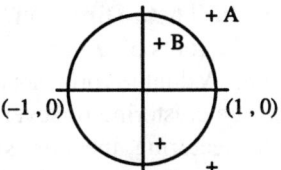

Figure 1. The Z–plane

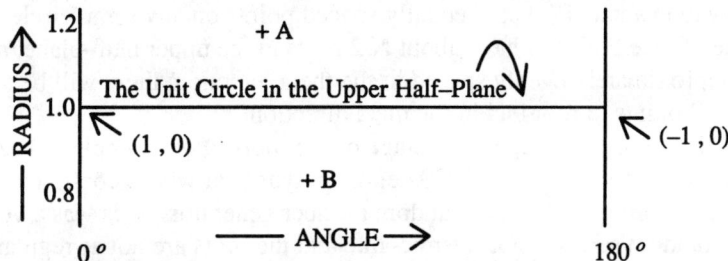

Figure 2. Rectangularized Upper Half of the Z–Plane

Figure 3 shows an example of the roots of a zero–phase wavelet. Note that the roots are in reciprocal pairs.

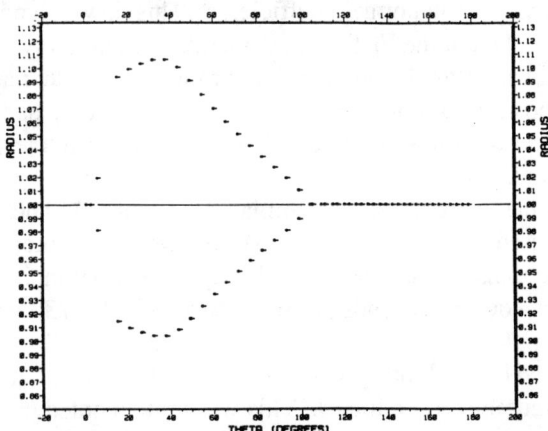

Figure 3. Root Plot of a Truncated Autocorrelation

4. Grid Searching for Roots. The real FFT provides an efficient means to evaluate a polynomial on a grid of points in the upper half–plane.

Consider the real–valued function $h(z) = |p(z)|$. If c is a local minimum of $h(z)$, then c must actually be a zero of $h(z)$ and $p(z)$ [2]. The strategy is to examine $h(z)$ at all 3x3 groups of neighboring points in the grid and test whether the value at the center is less than the values of the surrounding eight points. Any such minimum is assumed to be near a genuine root. The precision of this root estimate is dependent on the density determined by the FFT length as well as the exponential modulation increment used to move the FFT polynomial evaluations off the unit circle. If greater precision is required, Newton's method may be used to "polish" the root location and confirm the root's existence.

4.1 Creating a Dense Grid in the Theta Direction. As discussed in a later section, there is a very strong tendency for the roots of z–transforms of seismic–like data to be evenly spaced in the theta direction. A double root was never observed (except in untruncated autocorrelations) nor was a clustering of several roots with approximately the same theta coordinate. The regular sectoring, the almost nonexistence of multiple roots, and the fact that most roots lie very near the unit circle are the reason why grid search is effective.

If N is a power of 2 and P(z) is a polynomial of order N with real coefficients, we have seen how to evaluate P(z) at N equally spaced points on any semi–circle in the upper half–plane. Since P(z) will have about N/2 roots in the upper half–plane, and since they will be approximately evenly spaced in the theta direction, there will be about a 2 to 1 density of grid points to root loci in the theta direction.

This is sufficient to get a large percentage of the roots. For example, a 1024 point real FFT was used to try to factor a 1000 degree polynomial whose coefficients were generated by a uniformly distributed random number generator. This was a very severe test because a random series is not seismic–like and the roots are not as regularly spaced as the roots of a seismic–like z–transform. Also, some of the roots were much farther from the unit circle than normal. Nonetheless, 92% of the roots were found on the first pass. When a 1024 point FFT was used to factor a 500 degree polynomial with random coefficients, 96.6% of the roots were found; polynomial division found the rest.

The density of grid points in the theta direction can be doubled by appending N zeros to the end of the array of polynomial coefficients. This new polynomial will have exactly the same set of roots, but the FFT will be operating on an array that is twice as long and will therefore be evaluating the polynomial at roots of unity that are twice as dense. As a rule of thumb, the array of polynomial coefficients should be padded with sufficient zeros so that the number of points in the real FFT is at least twice the degree of the polynomial, preferably more.

In order to factor the hundreds of polynomials for Figures 11 and 12, a 1024 point real FFT was always used and usually 200 transforms were computed for radii ranging from 0.8 to 1.2. These transforms, together with the exponential multiplications and 3x3 point minima searches, took 30 seconds per trace on a VAX 11–780 with FPS–100 array processor.

When implemented on a 12 mhz. PC clone (Intel 286 with 287 coprocessor), this same level of processing took 7.5 minutes. This was implemented using double precision arithmetic in a fourth generation language written in C by one of the authors. The overhead of the language is not known; however, a single 1024 point FFT required 1.25 seconds, so 200 of them is slightly over 4 minutes. Also, several optimizations were available but were not tried, so the minimum time is somewhere between 4 and 7.5 minutes.

An 8192 point real FFT was sufficient to completely factor a 2000 degree polynomial. When the roots were polished and recombined, the resulting polynomial coefficients agreed with the original polynomial to 10 decimal places.

4.2 Creating a Dense Grid in the Radial Direction. Multiplication of the polynomial coefficients by the sequence

$$1 \ , \ r \ , \ r^2 \ , \ r^3 \ , \ ... \ , \ r^{N-1}$$

before the FFT computation, evaluates the polynomial on a semi–circle of radius r. Thus, it is easy to adjust the grid density in the radial direction by the choice of r.

This is important because, as you move away from the unit circle, the polynomial magnitude grows at the rate r^{N-1}. For example, one polynomial had a root with radius 1.4 (which was very abnormally large). The absolute value of the complex derivative at the root was 1.0E32. Double precision arithmetic was mandatory in most programs.

High degree polynomials change much more rapidly in the radial direction than in the theta direction. It is suspected that this is the reason why grid search occasionally missed roots that were located within the grid. A polynomial division step was required to locate them. Various schemes for reducing the effects of this astonishing slope in the radial direction were considered but never implemented.

Typically a search was performed from r = 0.80 to r = 1.2 in 200 steps; most roots of 250–500 degree polynomials lie in this range. The longer the polynomial, the more likely the roots would lie even closer to the unit circle and the search could be made over a smaller radial range. In fact, it is often easier to factor a 300 degree polynomial than a 150 degree polynomial!

5. Using Polynomial Division to Find the Remaining Roots. Usually grid search finds about 98% of all roots. For some applications, that is sufficient. If it is desirable to account for every root, further work is required.

To find the remaining roots, Newton's method is first used to polish the accuracy of root locations found by grid search. Convergence in 3 or 4 iterations to 15 significant digits is typical because the starting value is already close to the true root location. On occasion, a root will polish to a location far from the initial 3x3 point minimum to another 3x3 point minimum. In that case, two tentative root locations will polish to the same root. Therefore, the root file should be sorted by ascending values of theta and any duplicate roots eliminated. It is suspected that these false minima are caused by the amazingly large slope in the radial direction.

The resulting set of polished roots can be recombined by polynomial multiplication and the result divided into the original polynomial. The quotient will be a very short polynomial, degree 12 or less, since it contains only the roots missed in the primary search. This short polynomial can be factored by more conventional means.

For z–transforms that are not seismic–like, it is often the case that some roots will be unusually far from the unit circle. If a large number of the missing roots fall into this category, it may be impossible to factor such a polynomial by the grid search method.

6. Finding Missing Roots by Inspection. There is another method which can often find missing roots of seismic–like z–transforms. As mentioned earlier, there is a very strong tendency for such roots to be regularly spaced in the theta direction. If the roots found by grid search are plotted, one can often immediately spot the approximate position of a missing root by looking for a gap in the regular spacing! Newton's method will quickly find the exact position.

7. SOME GENERAL PROPERTIES OF ROOTS.

7.1 Quadratic and Linear Factors. Most polynomial roots are complex. Real roots do exist, but in far fewer numbers. Thus the basic building block for seismic–like data is the three–term real quadratic polynomial with complex conjugate roots.

7.2 Root Sectorization. Seismic–like (band–limited?) polynomial roots fall in strongly sectored positions in the theta direction. For example, a 180 degree polynomial will factor into 90 complex conjugate pairs, sectored into 2 degree sector widths ($180/90 = 2$). If there are real roots, there must be an even number since the polynomial order is even. This will not appreciably change the sector size but will possibly shift the sector positions about one degree in the theta direction.

Sectoring is stronger in the reject band than in the pass band. In fact, there may be an occasional empty sector at the very lowest end of the pass band.

The sector size for zero–phase wavelets is doubled since the roots must be in reciprocal pairs. This is illustrated in Figure 3 which is the root pattern of a truncated autocorrelation function of a seismic wavelet. If a symmetric wavelet is a true autocorrelation, any roots that fall on the unit circle are double roots. If a symmetric wavelet is not a true autocorrelation, as in Figure 3, the roots in the reject band are not multiple roots and separate into sectors with widths half those of the pass band reciprocal pairs. This is the typical pattern for truncated autocorrelations. Thus, truncation generally results in a symmetric wavelet that cannot be the exact autocorrelation of some other wavelet because the roots on the unit circle are not double roots.

7.3 Root Radii. The loci of root radii relative to the unit circle bears a striking resemblance to the Fourier amplitude spectrum. If the amplitude spectrum is zero at some frequency, the z–transform must have a root at an appropriate point on the unit circle to cause this. In order to get a "pass band," the roots must recede from the unit circle. Figure 3 is a good illustration of this.

8. EXAMPLES OF THE USE OF FACTORIZATION.

8.1 Phase Unwrapping [3]. It is very easy to unwrap the phase functions associated with linear and quadratic factors of a polynomial. The phase function of the original polynomial is the sum of the phase function of its factors. The only problem that can arise is that roots on the unit circle have a phase function with a jump discontinuity of 180 degrees.

8.2 Studies of Phase Effects. Figure 4 shows the roots of an autocorrelation which has been shifted so that the first nonzero sample occurs at time zero. Note the reciprocally paired roots. All roots lie off the unit circle with the exception of a root at about 87 degrees. This root must be a double root. Since this autocorrelation is not truncated, the phenomenon of double–density roots on the unit circle, as in Figure 3, is not seen.

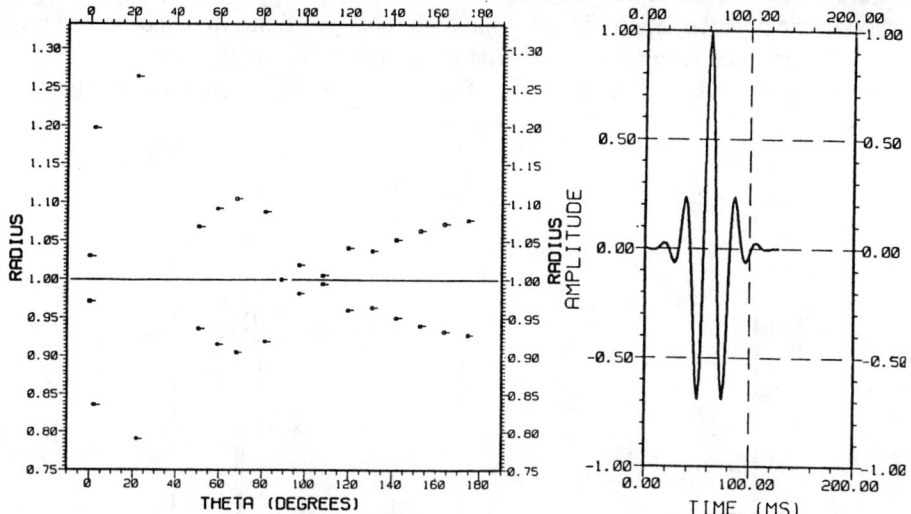

Figure 4. A Shifted Autocorrelation and Its Roots

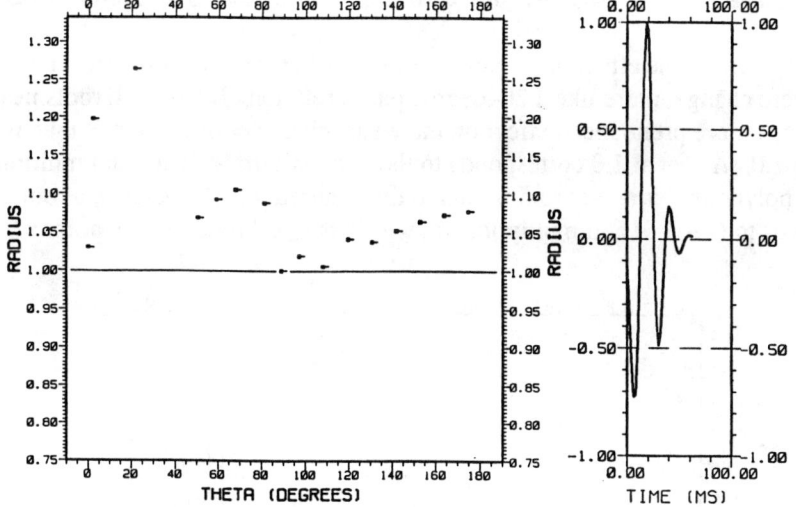

Figure 5. The Min–Phase Part of Figure 4

Figure 5 shows the min–phase half of the autocorrelation roots of Figure 4. The wavelet corresponding to these roots is also shown. Note the alternative method of determining the min–phase wavelet from its autocorrelation: 1) factor the autocorrelation, 2) select the minimum phase roots, 3) take any roots on the unit circle and move them outward by a tiny amount, and 4) polynomially recombine them. However, the polarity of the resulting wavelet will be indeterminate.

Step 3 in the procedure insures that the resulting wavelet is strictly min–phase. It is conceptually the equivalent of adding white noise to the autocorrelation matrix in the Wiener–Levinson method. However, in this approach, only the specific roots violating the rigid minimum phase requirement are adjusted. This step could be ignored with almost no visible effect on the resulting wavelet.

Figure 6 examines the effect on the wavelet of changing one of the roots in Figure 5 to a max–phase position, approximately the reciprocal position. In this case, the second root was chosen. The effect on the wavelet of changing this single root (and its complex conjugate) is essentially a slight time shift followed by a 180–degree phase rotation.

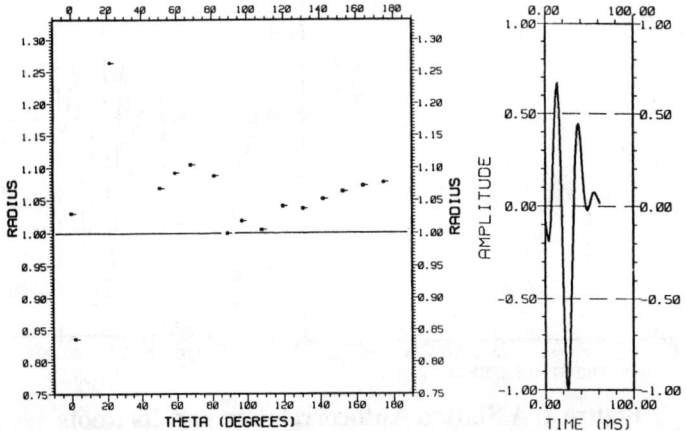

Figure 6. The Roots of Figure 5 but with the Second Root Switched to Max–Phase

Figure 7 shows the effect of a change of the fourth root to max–phase. The result is a less severe change, more like a 20–degree phase rotation. In general, roots nearer 1.0 have a much more pronounced effect on the visual character of a wavelet than roots farther from 1.0. A root at 1.0 corresponds to the z–transform $(z - 1.0)$ and multiplication with this polynomial is an approximation to differentiation. Conversely, a root at –1.0 corresponds to $(z + 1.0)$ and multiplication with this is 2.0 times a two–point running average.

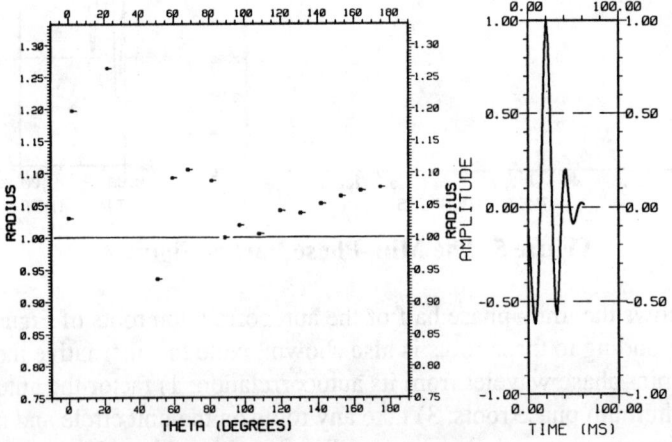

Figure 7. The Roots of Figure 5 but with the Fourth Root Switched to Max–Phase

Another experiment began with a truncated autocorrelation (not shown). All roots between 90 and 180 degrees were exactly on the unit circle. Figure 8 shows the roots of the associated min–phase wavelet generated by the Wiener–Levinson process. A white noise value of .01% has been added to stabilize the inversion. This is clearly seen to have caused a distinct reject band offset of all roots away from the unit circle. Pass–band roots are affected, but not as noticeably.

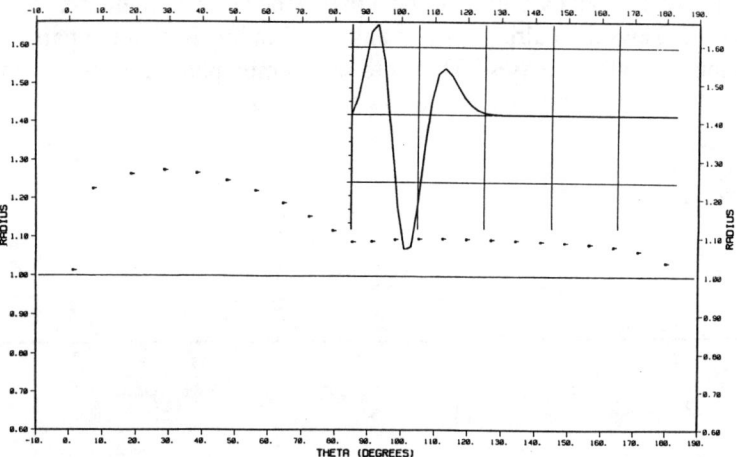

Figure 8. Roots of a Ricker Generated by Wiener–Levinson and .01% White Noise

While the autocorrelation factorization method for obtaining the min–phase wavelet is more accurate than the Wiener–Levinson method (because there is no necessity to add white noise), experience indicates that the differences in the two wavelets are small.

Figure 9 is a study of the roots of a strongly mixed–phase wavelet. The wavelet can be approximately corrected to even symmetry by simple phase rotation. Its root pattern shows it to be strongly mixed phase, with fewer max–phase roots than min–phase roots.

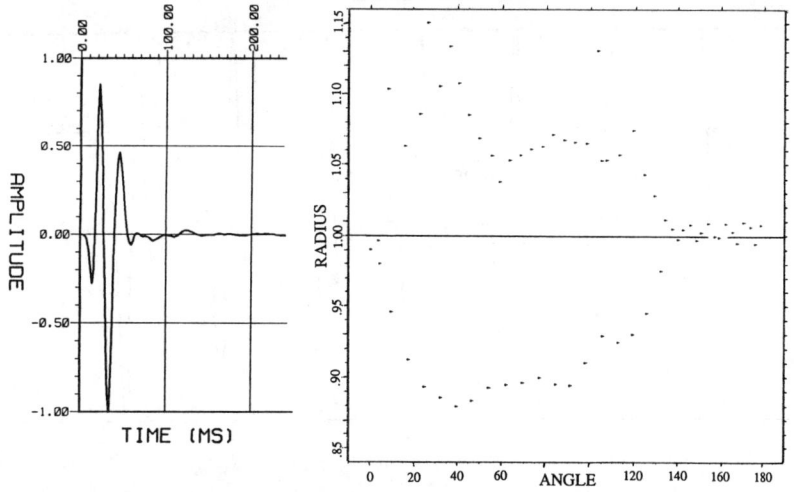

Figure 9. The Roots of a Strongly Mixed–Phase Wavelet

Figure 10 studies the effect of phase rotation on wavelet roots. The subject wavelet is a zero–phase Ricker which is rotated in 15 degree steps to 90 degrees. All root patterns are shown on one plot, coded by the direction of a pigtail attached to the root symbol. The original wavelet is shown with a pigtail pointing to the right. As phase rotation increases, the corresponding pigtails rotate counter–clockwise.

The min–phase roots (outside the unit circle) decrease in radius and shift toward lower angles while the max–phase roots increase in radius and shift toward larger angles. The original wavelet roots are in reciprocal pairs since it is zero–phase. The 90 degree shifted wavelet skews the pairing both in angle and radius so that the pairs come in alternating fashion as angle increases. That is to say, a min–phase root followed by a max–phase root followed by a min–phase ....

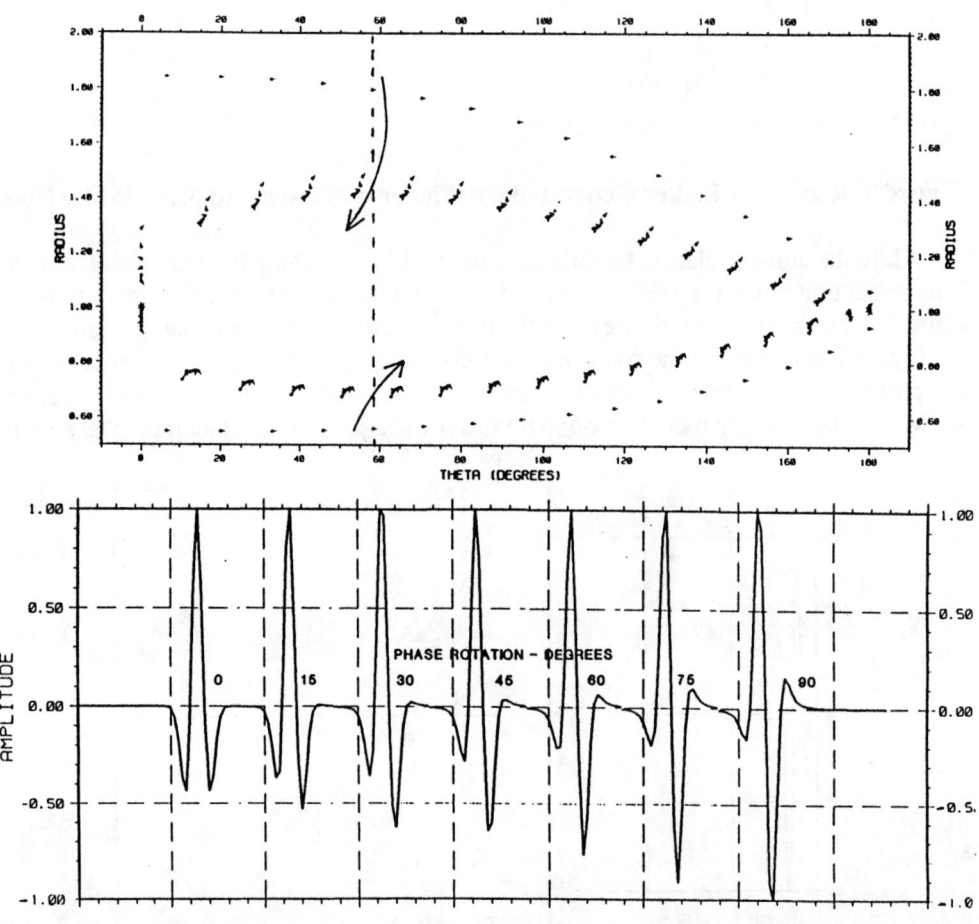

Figure 10. The Effect of Phase Rotation on Wavelet Roots

8.3 Testing the Convolutional Model. The convolutional model says that a seismic trace results from the convolution of a seismic–like, short wavelet with a much longer random series of spikes representing the reflectivity series. Under this model, if there were no truncation or noise, the roots of a seismic trace z–transform would contain a subset which represents the wavelet, the remainder would be attributable to the reflectivity function.

Truncation and noise causes unknown effects on the combined set of roots. However, there is some evidence that the wavelet roots continue to make their presence known, at least at low frequencies where the lithology spectrum is weakest.

Across a seismic section of many traces, the common wavelet roots should persist while the reflectivity roots should vary because of geologic factors (changing interval thickness and velocity, pinchouts, etc.) If one overplotted the roots obtained from a large number of seismic traces, there might be concentrations of certain root positions related to the seismic wavelet, which is invariant in the traces being factored. These clusters might not be actual wavelet roots because they will no doubt include the effects of a non–white lithologic generating function. However, the phase function taught by the clustered root positions probably are the dominant phase controlling roots for the wavelet plus the non–white lithology effect.

Figures 11 and 12 are overplotted root patterns for two such experiments. Figure 11 involved 191 traces, a 692 ms. window, and a 4 ms sample rate. The entire set represents a total of over 33,000 roots. Approximately half of them (the upper half–plane) are shown in Figure 11. About 98.5% of all the roots were found in the initial grid search. These were polished using Newton's method but a polynomial division step was not made to find the missing roots.

There is obvious root clustering in the lower end of the pass band where 10 to 12 clusters are seen. They are fairly well sectored and might be caused by wavelet roots. There was little else that was consistent from trace to trace; the geology was definitely changing.

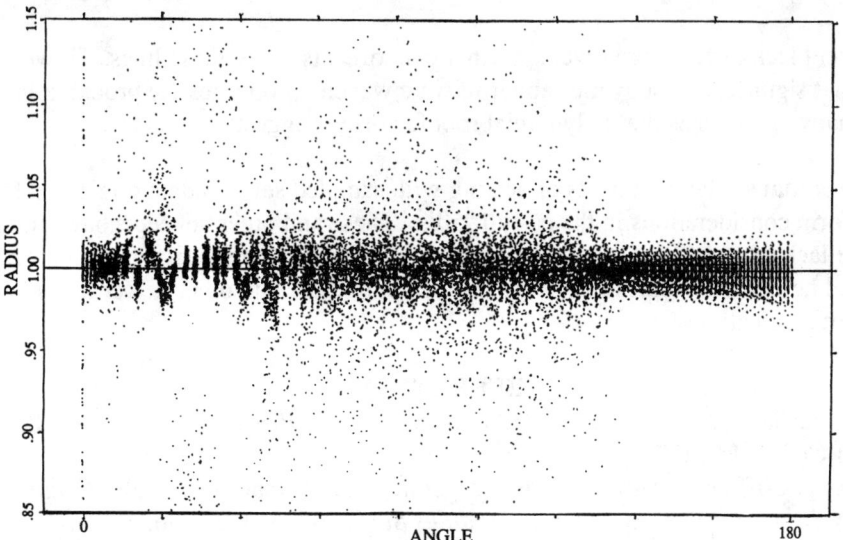

Figure 11. Polished Roots from a 692 ms Window on 191 Traces

Figure 12 is a second data set involving 101 traces of 852 ms. each. The roots are unpolished and clearly show the grid search pattern. This data set is not as high in signal–to–noise ratio as the first set and does not show the same degree of pronounced clustering. This condition was improved marginally by polishing the roots.

For each trace, an unwrapped phase function was computed from its roots. Then, for each section, the average unwrapped phase function was computed. The phase rotations implied by each of these two averages was used to "correct" the two sections. For the line associated with Figure 11, this procedure did a much better job of matching the synthetic than blindly rotating the phase by multiples of 90 degrees. For the line associated with Figure 12, this procedure did not help.

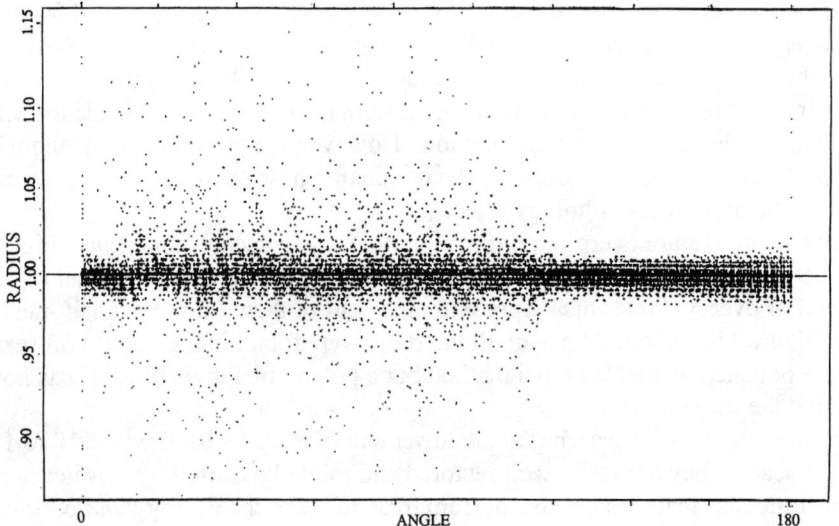

Figure 12. Unpolished Roots from a 852 ms Window on 101 Traces

Current lack of funding prevents further experiments along these lines. However, it seems that significant discoveries about seismic wavelet properties are probable in view of the many approaches that polynomial root factoring suggests.

9. Summary. Many methods of dealing with sampled seismic data originated from z–transform considerations in the early 1950's. These and other concepts have suffered from the lack of a viable root solving method. It is hoped that the factorization method discussed herein will result in a revival of interest in these and other newer ideas for z–transform applications.

REFERENCES

1. Oppenheim, A.V. and Shafer, R.W., *Digital Signal Processing,* Prentice–Hall, Englewood Cliffs, N.J., 1975.

2. Long, Cliff and Hearn, Thomas, The College Mathematics Journal, *"Graphing the Complex Zeros of Polynomials Using Modulus Surfaces"* , Vol. 20, No. 2, March, 1989.

3. Steiglitz, Kenneth and Dickinson, Bradley, IEEE Transactions on Acoustics, Speech and Signal Processing, *"Phase Unwrapping by Factorization,"* Vol. ASSP–30, No. 6, December 1982.

CHAPTER 8

Wavelet Estimation by Exponential Decay Method

M. T. Taner*
L. Lu**

ABSTRACT

Despite their intricate designs, seismic sources do not produce a
sharp, spike-like pulse. To achieve accurate imaging of the earth's
subsurface, estimating and shaping seismic source wavelets into a desired form is necessary. An autoregressive model (ARMA) is used for
seismic trace and the exponential decay method is addressed and
implemented for seismic wavelet estimation. A set of procedures is
outlined to overcome the difficulties arising from reverberations, low
signal-to-noise ratios and non-minimum phase characteristics. An
actual data example is presented to illustrate these problems and
their solutions. The practical aspects of wavelet shaping using an
ideal wavelet is also discussed.

INTRODUCTION

The objective of seismic exploration is to obtain accurate responses
of reflectors around deep target zones. This can be accomplished
by generating and processing data so that distortions caused by in-
struments or shallower horizons are suppressed or eliminated. One
distortion generated at the surface is the variation of seismic source
signature. This variation may result from many factors, such as surface conditions, source coupling, shot size, source type and depth.
Without signature corrections, correct reflector responses cannot be
obtained and proper stratigraphic interpretation cannot be achieved.

*Seismic Research Corporation and Rice University.
**CogniSeis Development, Inc.

A number of wavelet deconvolution, source signature (wavelet) estimation and shaping methods exist in industry. For example, to derive the wavelet spectral estimate, Potts et al. (1982) use the minimum-phase assumption and the spatial averaging of amplitude spectra. Potts also uses these methods in designing a two-sided inverse filter for wavelet correction. Lindsey (1988) calculates and interprets the roots of wavelet polynomials obtained from the autocorrelation function. Wiggins (1978) discusses the minimum entropy deconvolution method, which makes no assumptions regarding the phase of the source wavelet. Jurkevics and Wiggins (1984) evaluate several algorithms for prediction error filters and algorithms which do not assume a minimum-phase condition, e.g., minimum entropy, homomorphic and zero-phase deconvolutions. They conclude that standard Wiener prediction deconvolution is robust under many input conditions. Jurkevics and Wiggins also conclude that substantial improvements are obtained by Burg's algorithm for short data segments and the minimum entropy method for mixed phase wavelets with sparse reflection series.

In this paper, we would like to address a stable and robust method for wavelet estimation: the exponential decay method. We will outline a set of procedures designed to overcome the difficulties arising from reverberations, low signal-to-noise ratios and non-minimum phase characteristics. Practical aspects of wavelet shaping using an ideal wavelet approach will also be discussed.

SOURCE SIGNATURE

Compared to marine recording, field recording of the source on land is somewhat complicated and therefore is not generally practiced. The seismic source signature is usually recorded in marine acquisitions. In some instances, this is essential; otherwise, records would be useless. In particular, Maxipulse*, Vaporchoc** and water gun require recorded signatures because of their complex signature shape and their shot-to-shot variation. Other marine sources, such as airgun and sleeve exploder, are not usually recorded. Even though their shapes are not far from ideal, these marine sources should be also recorded.

* Trade Mark of Western Geophysical.
** Trade Mark of C.G.G.

If the source signature is measured before the survey or recorded with the shot, we may design a correction filter which can be used for shaping the wavelet to a desired output. This filter may then be applied to the data. If the source signature is not recorded during data acquisition, the first step in correcting it is to estimate (or extract) it directly from the data. Since the estimation process involves computing an average characteristic function of the common source data, successful results will depend largely on the signal-to-noise ratio. Should the data contain significant ground-roll, a pre-processing step to suppress or eliminate this factor is necessary.

THE EXPONENTIAL DECAY METHOD

The basic assumptions of the exponential decay method are listed below:

1. Seismic wavelet (signature) is causal (one-sided). This assumption is generally valid since almost all seismic energy sources are causal, i.e., energy is transmitted only at and after the time of detonation, not before.

2. Seismic signature after detonation will generally decay with time. Generally, these causal signatures contain primarily minimum phase components. Note that we emphasize the causality as the general assumption, not minimum phase. Vibroseis signals which transmit energy without decay over long time periods (20 seconds or longer) are exceptions.

3. Seismic trace is defined by the autoregressive and convolutional (ARMA) model. The seismic source signature is assumed to be a short pulse convolved by the reverberant component (Fig. 1). This component is generated by the coupling action between the energy source and the earth. The ARMA model for a seismic trace can be expressed as:

$$f(t) = r(t) * s(t) * d^{-1}(t) + n(t) \qquad (0.1)$$

where:

$$
\begin{aligned}
f(t) &= \text{seismic trace} \\
r(t) &= \text{reflectivity series} \\
s(t) &= \text{seismic source wavelet} \\
d^{-1}(t) &= \text{reverberation} \\
d(t) &= \text{a short dereverberation (or} \\
&\quad \text{deconvolution) operator} \\
n(t) &= \text{noise}
\end{aligned}
$$

We assume that the wavelet $s(t)$ is a mixed-phase function and $d^{-1}(t)$ is minimum-phase (Deeming 1989). Thus, a simple, short deconvolution operator $d(t)$ should be sufficient to correct the wavelet reverberation.

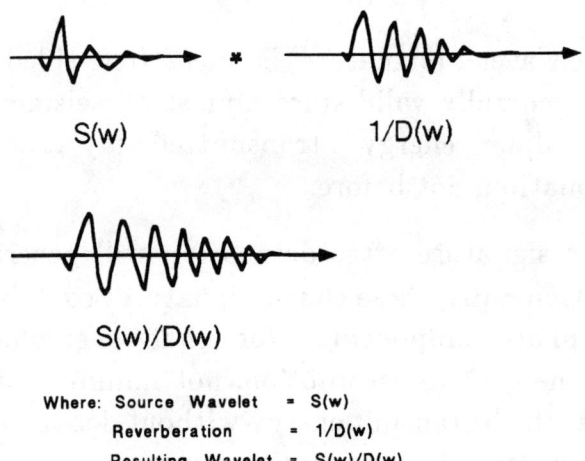

S(w)   1/D(w)

S(w)/D(w)

Where: Source Wavelet = S(w)
       Reverberation   = 1/D(w)
       Resulting Wavelet = S(w)/D(w)

Fig. 1.   The autoregressive and convolutional model.

4. The receiver array responses are nearly constant with negligible variations along the seismic line. If not, the source and receiver functions can be computed in a surface consistent manner (Taner and Coburn 1980).

5. For common source gathers, we assume the noise $n(t, x)$ and the reflectivity functions vary from trace to trace. Therefore, the average amplitude spectrum of these traces represents a good estimate of the amplitude spectrum of the source signature.

Taner and Coburn (1980) proposed the exponential decay method for seismic wavelet estimation. This concept was based on the fact that mixed phase wavelets can be converted to minimum phase by an appropriate exponential scaling (Schafer 1969, and Oppenheim and Schafer 1975). Let the source signature $S(z)$ be represented by the Z-transform or the polymonial with N roots z_n, $n = 1, 2, \ldots, N$ as:

$$S(z) = s_0 + s_1\, z^1 + - - - + s_N\, z^N = s_N\, \prod_{n=1}^{N} (z - z_n) \quad (0.2)$$

Where:

z	$= e^{(iw\, dt)}$ unit delay operator
dt	$=$ sampling interval time
z_n	$= |z_n|\, e^{i\theta_n}$
$|z_n|$	$=$ the modulus
θ_n	$=$ the phase of the nth root

By definition, if $S(z)$ is minimum phase, all roots must lie outside the unit circle on the complex plane, $|z_n| > 1.0$. If some of the roots $|z_n| < 1.0$, then the corresponding wavelet is mixed-phase. Most seismic source signatures are mixed-phase, with the majority of their roots lying outside the unit circle.

The source signature can be made minimum phase by substituting z with rz in the above formula, where r is a value less than or equal to 1.

$$S(rz) = s_n\, r^n \prod_{n=1}^{N} (z - z_n/r) \quad (0.3)$$

This equation shows that the roots of the new polynomial $S(rz)$ are z_n/r. If r is a positive real number, the corresponding modulus becomes $|z_n|/r$. An example of the source wavelet $S(z)$ and the exponential scaled source wavelet $S(rz)$ is shown in Fig. 2. Let us now consider that some roots of $S(z)$ lie inside the unit circle. If we select $r < \min |z_n|$, then $|z_n|/r$ will be larger than one, hence; all the roots will lie outside the unit circle and $S(rz)$ becomes the minimum phase polynomial.

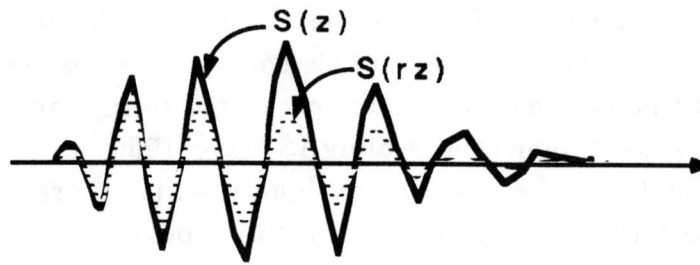

Fig. 2.   The source wavelet S(z) and the exponentially scaled source
          wavelet S(rz).

Once the trace is exponentially scaled according to the radius r, the minimum phase wavelet can be computed by conventional methods. Let $B(z)$ be the computed minimum phase wavelet. The mixed-phase wavelet $S(z)$ can be easily obtained by inverse scaling as:

$$S(z) = b_0 + (b_1/r)z + - - - - +(b_n/r^n)z^n = B(z/r) \qquad (0.4)$$

IMPLEMENTATION

The exponential decay method is a stable and robust method with implementation of the procedures outlined below:

First, we use the ARMA model in which the seismic wavelets may be simplified or shortened by a deringing operator. This deringing operator can be a very short decon operator as described in the literature (e.g., Deeming 1989). If this process (the inverse of an autoregressive process) is not applied, the wavelet shaping cannot be accomplished successfully. Lazear (1984), based on a simple convolutional model, concludes that the exponential method is useful only for seismic wavelet estimation when data windows may be chosen such that no wavelets are truncated at the window's onset.

In our estimation, most other methods have similar problems in deriving accurate autocorrelation or amplitude spectrum. However, applying the simple deringing filter to the input data before wavelet estimation will reduce the possibility of wavelet truncation and will eliminate the disturbing poles from the complex plane. Most energy sources are near minimum phase and the exponential decay method is generally applicable to data collected with these sources. For cases of highly non-minimum phase pulses, such as Maxipulse* or Vaporchoc**, wavelets are usually recorded with the shot. Furthermore, they can be efficiently corrected using the deterministic approach, rather than using any wavelet estimation approach.

Second, the estimate wavelet amplitude spectrum can be improved by locating the beginning of the window at the local power minimum (or envelope), as indicated in Fig. 3. Time gates not starting at the local power minimum will certainly increase the possibility of wavelet truncation. Multi-space and time gates can be used to improve the

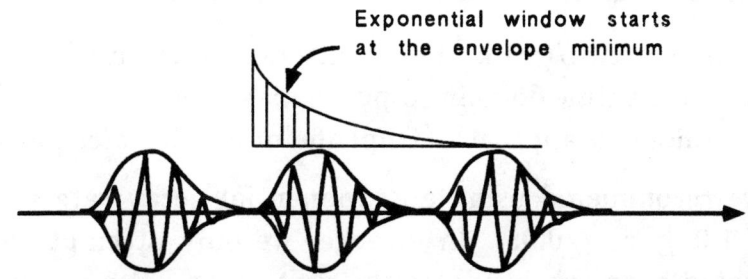

Exponential window starts
at the envelope minimum

Fig. 3.   An example of exponential window.

signal-to-noise ratio for estimating the wavelet spectrum. The following schemes can be used for averaging wavelet spectra obtained from the time and space gates:

Let:
$S_n(w)$ = the estimated amplitude spectrum of the source wavelet.
$A_n(w)$ = the amplitude spectrum of data in the nth window.
N = the total number of windows.

1) Linear averaging:

$$S^2(w) = \sum_{n=1}^{N} A_n^2(w)/N \qquad (0.5)$$

2) Harmonic averaging:

$$S^2(w) = exp\left[\sum_{n=1}^{N} ln\, A_n(w)/N\right] \qquad (0.6)$$

3) Average cross-power spectra:

$$S^2(w) = \frac{\left[\left(\sum_{n=1}^{N} A_n(w)\right)^2 - \sum_{n=1}^{N} A_n^2(w)\right]}{N[N-1]} \qquad (0.7)$$

The average of all pairwise cross-power spectra between all computational windows is computed. This is the most suitable way to compute the average wavelet spectrum because it actually corresponds to semblance computation in the frequency domain.

Since the estimated $S(w)$ is the amplitude spectrum of a minimum phase wavelet, its time domain response may be easily obtained using the cepstral domain approach (Oppenheim and Schafer 1975).

Third, we recommemd using an exponential decay rate r over the range of $1.0 \geq r \geq 0.95$. In essence, we only attempt to resolve roots located near the inside of the unit circle. For $r = 0.95$, the amplitude at 100th sample later in time will be 0.06 times smaller: $\left(\frac{1}{.95}\right)^{100} = .006$. The very large decay rate, associated with small r, would cause computational problems, which in turn, would cause an unstable process.

Generally, the decay rate r can be determined as described below. An analysis can be done by estimating the wavelets with different r decay values. The wavelets are estimated at Δr decrements of 0.001 or 0.002. The cross-correlation coefficient between consecutive estimated wavelets is computed to measure the degree of estimated wavelet similarity.

Suppose the cross-correlation coefficient between decay rates r_1 and $r_1 - \Delta r$ is less than 1. This indicates that the wavelets are not identical and that the root distributions for these two decay ratios are different, i.e., some roots between the radii r_1 and $r_1 - \Delta r$ exist on the complex plane. Hence, we try the new decay rate $r_1 - 2\Delta r$, and so on. If the cross-correlation coefficient is equal to 1.0 for decay rates r_k and $r_k - \Delta r$, then the estimated wavelets are identical for both rates; thus, no roots exist within the circles represented by decay rates r_k and $r_k - \Delta r$. This also indicates that all roots are now outside the radius r_k, making it the desired rate for wavelet computation. The estimated wavelets will remain the same for all decay rates less than

r_k (within a reasonable range and computational accuracy). For this reason, we do not need to know the exact decay rate.

A reasonable assumption is that r may change from shot to shot. To establish the decay rate's lower boundary (for production processing), we must analyze several typical shots along a given line. If large variations exist, we may divide the lines into proper segments and use the decay rate appropriate for each segment.

WAVELET SHAPING

The shaping of source wavelets first requires the definition of a desired wavelet shape, i.e., its band limits and whether it is zero or minimum phase. In practice, we generally use half of the folding (aliasing) frequency as the upper limit. At this frequency, we have four samples per cycle, which is sufficient to yield accurate measurements for computing both amplitude and phase. For the lower limit, generally the inverse of twice the recording length, or the low cut frequency employed in field recording, is used.

Several researchers have investigated the problem of the "ideal response function" (Berkhout 1974 and Schoenberger 1974). The general conclusion is that zero-phase signatures provide the best resolution. Reilly et al. (1979) discuss the desired shape of the amplitude spectrum. They conclude that it is possible to produce stable filters capable of great attenuation without harmful side-effects. These filters have continuous frequency domain responses, with continuous derivatives, and do not exhibit ringing or Gibbs phenomenon. They offer the geophysicist great flexibility in selecting the correct filter to enhance data and improve the signal-to-noise ratio. The interesting aspects of this type of filter will be reviewed below.

Let us now define a more resonable class of "ideal" filters, based on performance:

- Wide frequency band with smooth transition zones into reject bands.

- Zero-phase spectrum.

- Continuous amplitude spectrum without discontinuities of its derivatives.

- Time response and its envelope vary smoothly.

- Its envelope in the time domain decays at an exponential rate.

In general, filters suppress all frequency bands by some amount. However, by definition, the pass band is where the suppression is less than half power. The amplitude spectrum of the proposed filters $A(f)$ in frequency domain is given by:

$$lnA(x) = B_L \, e^{\beta_L(x_L - x)} + B_H \, e^{\beta_H(x - x_H)} \qquad (0.8)$$

Where:

$x = log_2 f$, (frequency given in octaves)
B_L, B_H, β_L and β_H parameters to be determined by the user given (x_H, x_L) high and low cut points and roll-off rates in db (octaves).

An example of the time and frequency domain responses of this filter type is shown in Fig. 5. The main peak-to-side lobe ratio is higher than other filters, such as the box car, Ormsby or Butterworth. Its envelope shows smooth, continuous decay. However, the amplitude spectrum is not flat-topped like that of the Butterworth filter. The time domain response shows rather uncomplicated and small side lobes and therefore has minimum interference with adjacent events. The sharp decay of the envelope, as well as the time response, improve event definition, event separation and sectional clarity of seismic data. Once the desired wavelet is determined, a two-sided Wiener inverse (or shaping) filter can be designed (Wood et al. 1978).

EXAMPLE

The above exponential decay method for wavelet estimation and the wavelet shaping approach is a robust and stable procedure which has been used for wavelet processing in industry for a number of years. An example of a land data set with two field records (Fig. 4) is selected to illustrate the wavelet estimation and shaping process. All data shown in this paper is automatic gain recovery applied for display purposes.

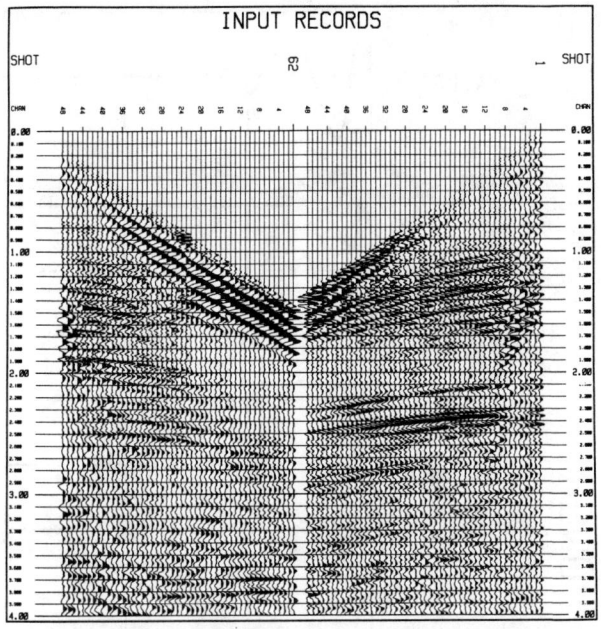

Fig. 4.   The input shot records for wavelet estimation.

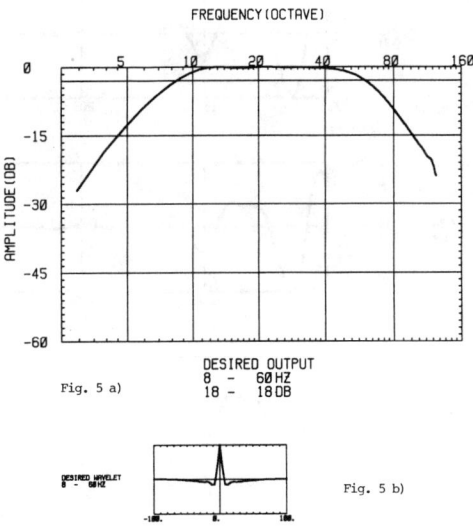

Fig. 5.   a) The spectrum of the desired wavelet.
           b) The desired wavelet.

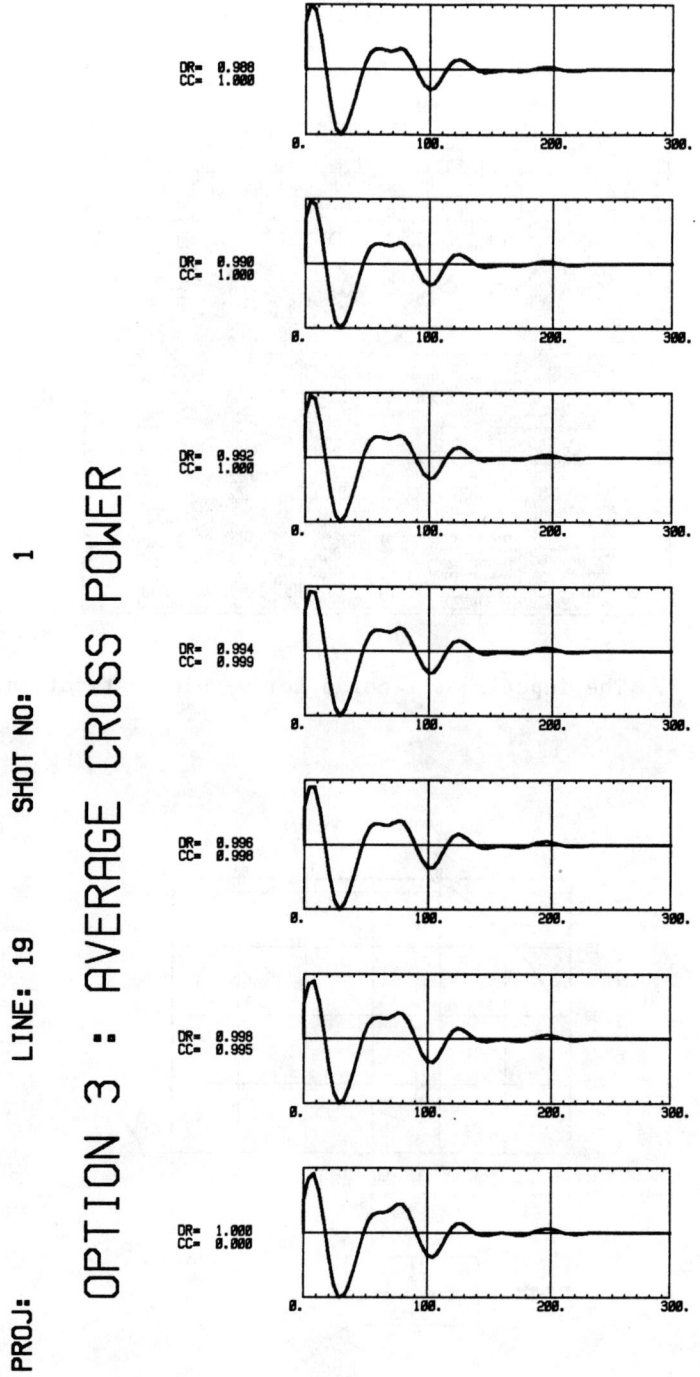

Fig. 6.   The computed wavelets at varying decay rates.

Before the wavelet estimation, a filter analysis is performed to determine the coherent signal bandwidth. Shot 1 has a wider band signature than shot 62 which has a more dominant low frequency reverberation. The peak wavelet spectra is approximately 15-16 Hz for shot 1 and approximately 9 Hz for shot 62. Since the data time sampling was 4 ms, we have chosen a desired output of 8-60 Hz band width with 18 db/octave roll-off on either end. The time and frequency domain responses of the desired outout wavelet are shown in Fig. 5.

We will first demonstrate the effect of the wavelet estimation without minimum inverse deconvolution for reverberation suppression. The average cross power option was selected for averaging the spectra obtained from varying time and space gates chosen to cover the main signal zone of the input data set. The estimated wavelets with differing decay rates are shown in Fig. 6. The decay rates for shot 1 are shown from $DR=1.0$ (with no decay) to $DR=0.988$ at decrements of 0.002. The notation CC represents the cross-correlation coefficients between adjacent computed wavelets.

The first computed wavelet with $DR=1.0$ is shown at the bottom of the figure. The second wavelet has $CC=0.995$ which indicates that the first and second wavelets are different. This suggests that some zeroes exist between the unit circle and a radius 0.998. The cross-correlation coefficient is equal to 1 between wavelets with $DR=0.994$ and $DR=0.992$, meaning that these two wavelet estimates are identical. Thus, all zeroes are outside the 0.994 radius circle. Note that all cross-correlation coefficients with decay rates smaller than 0.992 remain at unity, indicating that the wavelet estimation is stable and no further zeroes are encountered, as expected. In production processing, we therefore use a decay rate of 0.994.

In Fig. 7, the accuracy of wavelet shaping is tested. The wavelet with decay rate $DR=0.994$ is used. The shaping filter is displayed above the chosen wavelet. The filter convolved with the estimated wavelet will produce the actual output. Amplitude spectra of the computed wavelet and its shaped actual output are displayed with amplitude in db and frequency in octave scale. We have used an estimated wavelet length of 300 ms. Obviously, this length is inappropriate since the ringing wavelet theoretically is infinite in length, practically it is definitely much longer than 300 ms. Thus, using a shorter wavelet length would cause errors in wavelet estimation.

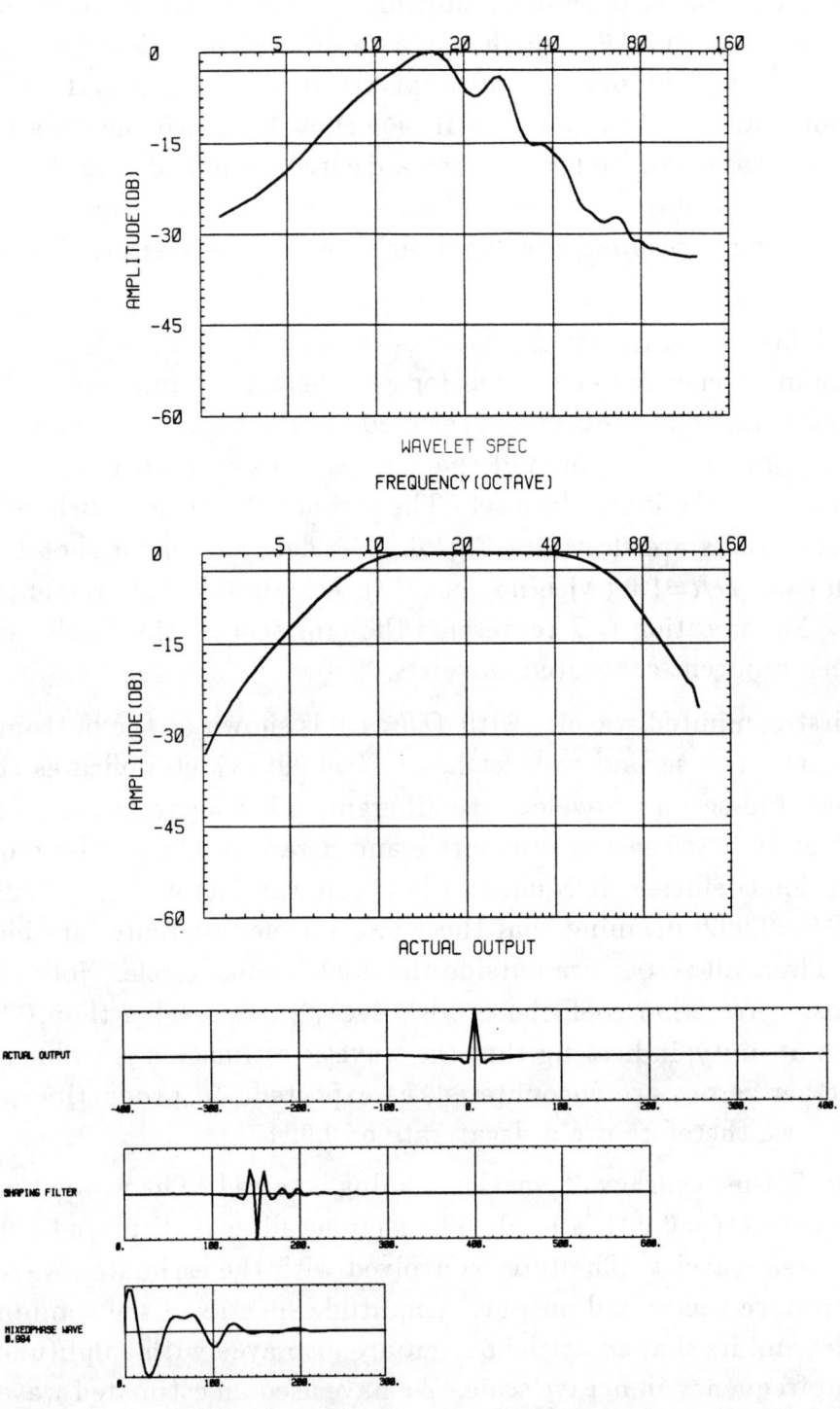

Fig. 7. A computed wavelet with its shaping filter, actual output,
wavelet spectrum and actual output spectrum.

On the other hand, computer word size and inaccuracies arising from the exponential decay application do not allow us to precisely estimate wavelets longer than 400 ms. Therefore, using an ARMA model handling the reverberant pattern separately is a more practical alternative. Although the shaping operation has performed satisfactorily, the estimated wavelet spectra of shot 1 (Fig. 7) shows amplitude variations in excess of 30 db. The seismic data after wavelet shaping without deconvolution is shown in Fig. 8.

A minimum-phase inverse deconvolution filter 60 ms in length is designed and applied to the input data. The deconvolved result indicates a relative increase of the high frequency content (Fig. 9). This deconvolution is the inverse of the reverberation caused by the source and earth coupling during the initiation of the seismic wave-field. Fig. 10 shows the analysis performed on shot 1. Again, an approximate decay rate of 0.994 produces stable wavelets. Since minimum phase inverse filtering alters only the reverberation, causing poles, it does not alter the zeroes of the wavelet's convolutional part; hence, the same decay rate is obtained. The shaping filter, final output and wavelet spectra are shown in Fig. 11. The wavelet shaping filter is then applied to the deconvolved data (Fig. 9) to produce the wavelet shaped data in Fig. 12.

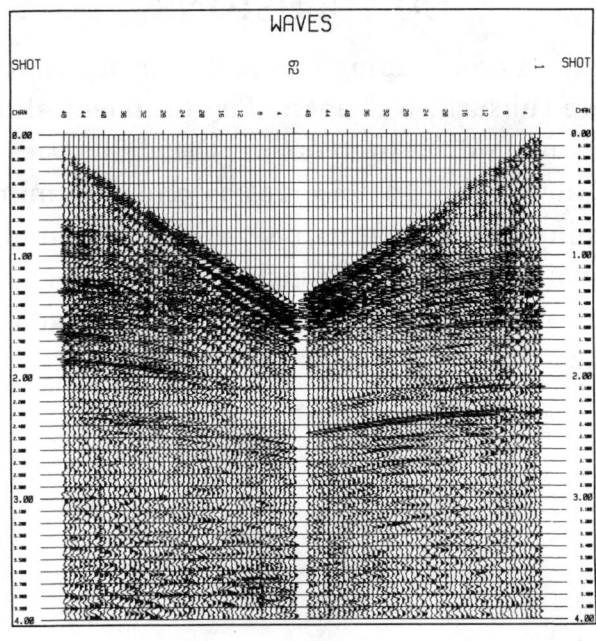

Fig. 8. The shot records (without decon) after wavelet shaping.

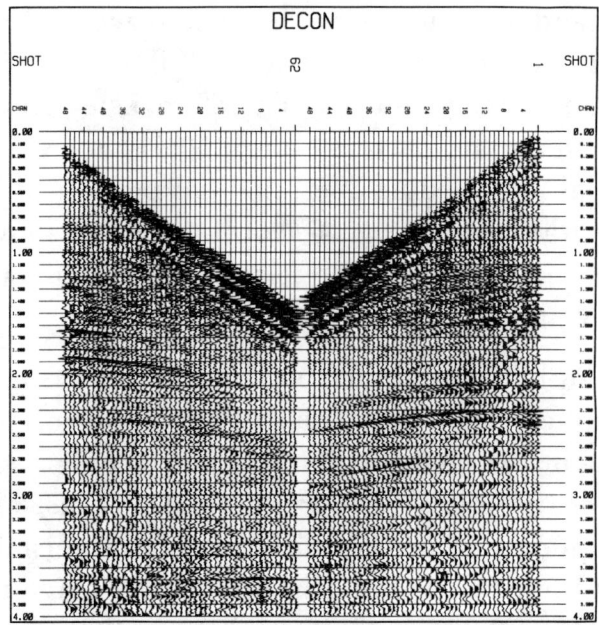

Fig. 9.   The shot records after decon.

By comparing the input (Fig. 4), the deconvolved (Fig. 9) and the wavelet shaped data (Fig. 12) (2.0 to 2.5 sec. on shot 1 and 1.5 to 2.5 sec. on shot 62), we see that events are improved and lateral coherency is enhanced by wavelet shaping.

CONCLUSIONS

Wavelet estimation and shaping is essential in stratigraphic processing for accurate subsurface images. The exponential decay method is an effective method in overcoming the difficulties arising from reverberation, low signal-to-noise ratios and non-mimimum phase characteristics in wavelet estimation. An ideal desired filter output for wavelet shaping has been defined and implemented. The above wavelet estimation and shaping process is a stable and robust method that has been used in industry for the last ten years.

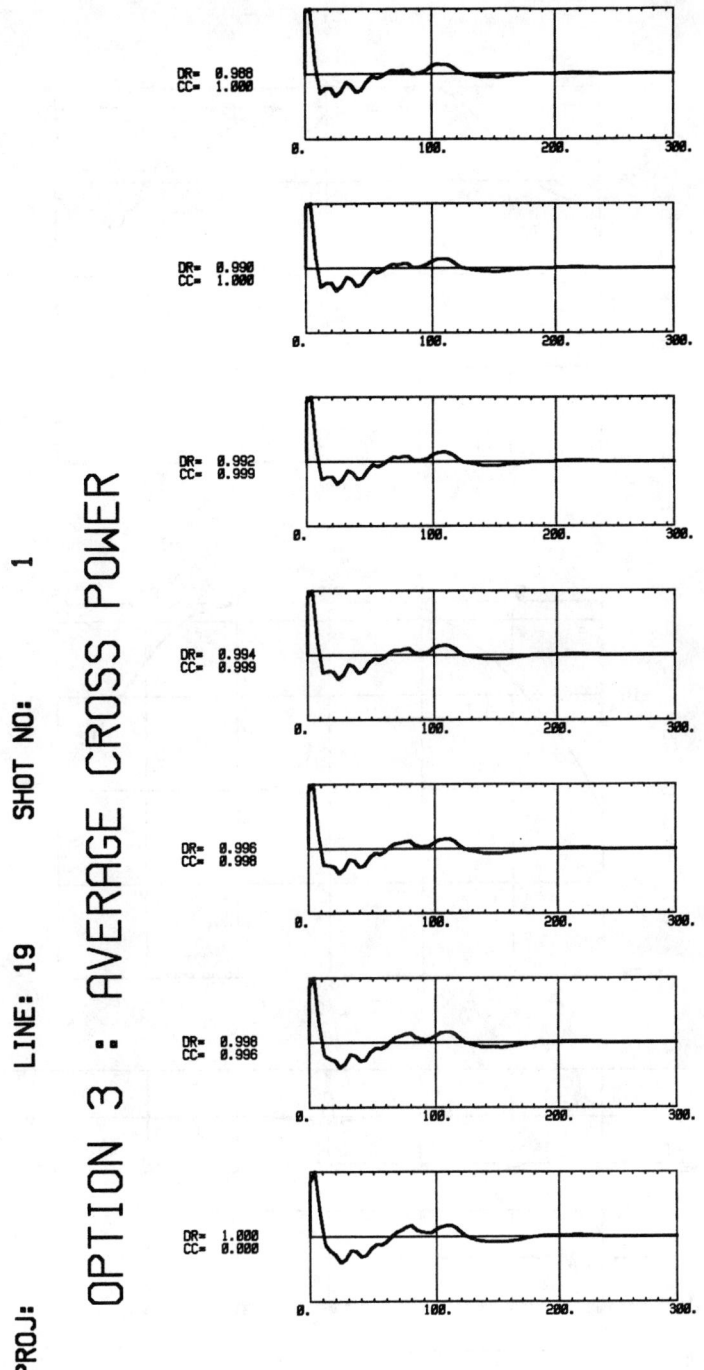

Fig. 10. The computed wavelets (from input with decon) at varing decay rates.

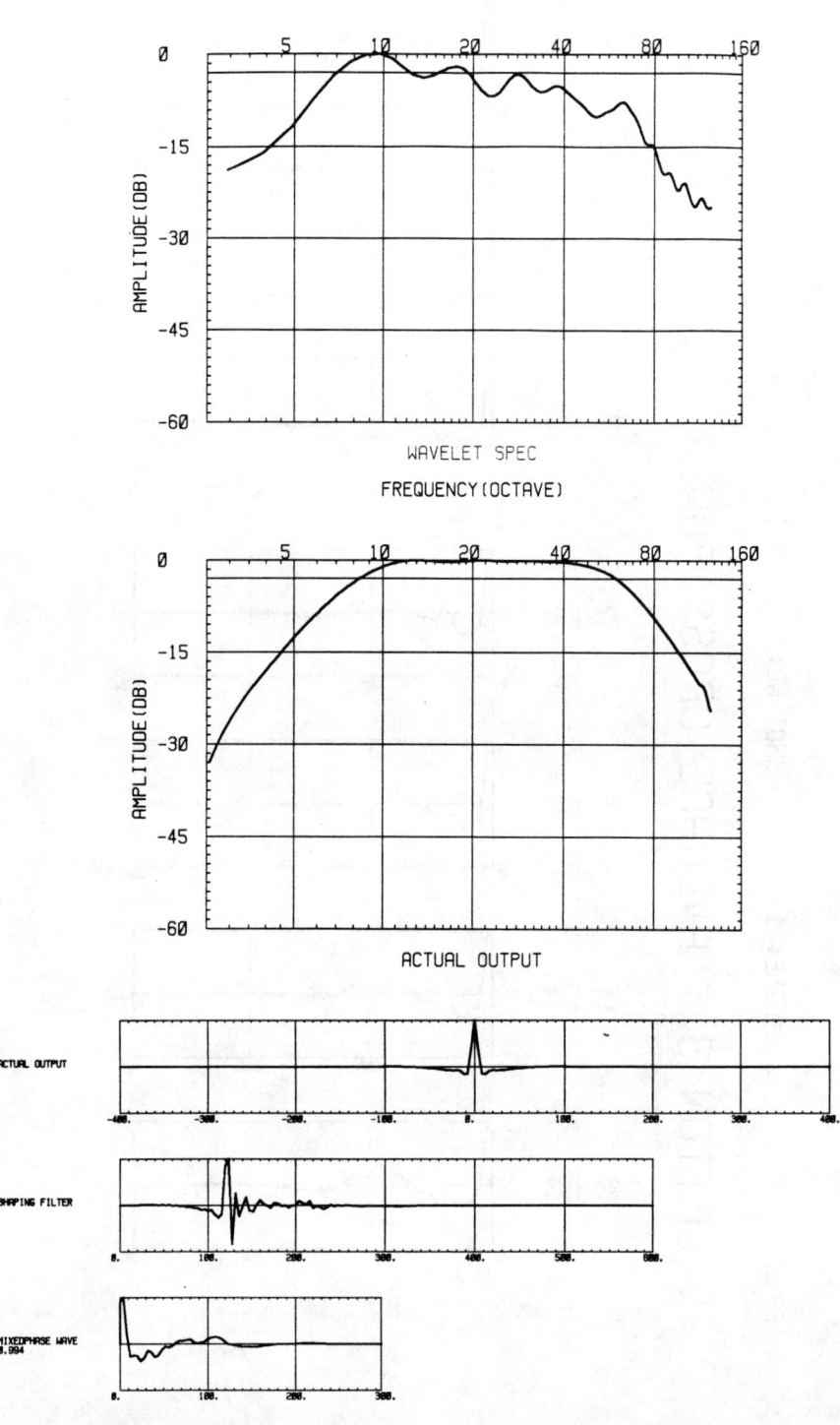

Fig. 11. A computed wavelet (from input with decon) with its shaping filter, actual output, wavelet spectrum and actual output spectrum.

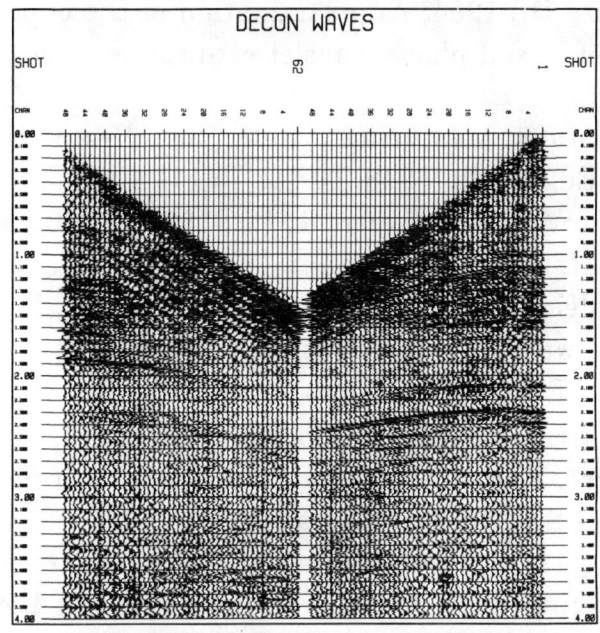

Fig. 12. The shot records (with decon) after wavelet shaping.

ACKNOWLEDGMENTS

The exponential decay method was first proposed by Dr. Fulton Koehler in 1976, which was later incorporated into the WAVES©(wavelet estimation and shaping) program with modifications indicated in this paper.

The authors wish to thank Seismic Research Corporation for allowing the publication of this paper which contains the basic design logic for the WAVES©program. They also wish to thank Ms. Suna Purser of CogniSeis Development, Inc. for editing the manuscript.

REFERENCES

Berkhout, A. J., 1974, Related properties of minimum-phase and zero-phase time functions: Geophysical Prospecting, **22**, 683-709.

Deeming, T. J., 1989, The physics of minimum phase, 51st Meeting of the European Association Exploration Geophysicists, West Berlin.

Jurkevics, A. and Wiggins, R., 1984, A critique of seismic deconvolution methods: Geophysics, **49**, 2109-2116.

Lazear, Gregory D., 1984, An examination of the exponential decay method of mixed-phase wavelet estimation: Geophysics, **49**, 2094-2099.

Lindsey, J. P., 1988, Measuring wavelet phase from seismic data: The Leading Edge, July 1988, 10-16.

Oppenheim, A. V., and Schafer, R. W., 1975, Digital signal processing: Prentice-Hall.

Potts, M., Schleicher, K., Wason, C., and Ellender, S., 1982, Pre-stack wavelet deconvolution, Presented at the 52nd Annual Meeting, SEG, Dallas, Texas.

Reilly, M.D., Greer, L.A., and Coburn, K.W., 1979, What is an ideal filter?, Presented at the 49th Annual International Meeting of S.E.G., New Orleans.

Schafer, R. W., 1969, Echo removal by discrete generalized linear filtering: M.I.T. Research Laboratory of Electronics, Technical Report No. 466, Cambridge, Massachussetts.

Schoenberger, M., 1974, Resolution comparison of minimum-phase and zero-phase signals: Geophysics, **39**, 826-833.

Taner, M. T., and Coburn, K. W., 1980, Surface consistent estimation of source-and-receiver response functions, Presented at 1980 Annual Meeting of SEG, Houston, Texas.

Wiggins, R., 1978, Minimum entropy deconvolution: Geoexpl., 16, 21.

Wood, L. C. Heiser, R. C., Treital, S., and Riley, P. L., 1978, The debubbling of marine source signatures: Geophysics, **43**, 715-729.

CHAPTER 9

Eulerian-Lagrangian Least Squares Collocation in Multi-Spatial Dimensions*

L. R. Bentley**

G. F. Pinder†

Abstract. Eulerian-Lagrangian least squares collocation (ELLESCO) is an accurate numerical method for solving the advection-dominated transport equation. ELLESCO uses the method of characteristics to track collocation points to the previous time level. The time derivative is approximated by a finite difference along the characteristic of the first order hyperbolic portion of the operator, eliminating the advection terms. Space is discretized with finite elements and cubic Hermite basis functions, and the algebraic system of equations is formed using least squares collocation. Numerical results from two dimensional test problems are used to compare ELLESCO performance to that of established Eulerian-Lagrangian methods (ELMs). Using twice the grid spacing, ELLESCO can match the accuracy of the other ELMs.

1. **INTRODUCTION.** Classical finite element and finite difference solutions to the advective-dispersive transport equation are contaminated by serious phase and amplitude errors when the advection term becomes large [1]. Methods known as "Eulerian-Lagrangian Methods" (ELMs) or "Modified Method of Characteristics" (MMOC) have been developed to improve the behaviour of numerical solutions when advection dominates [2,3,4,5,6,7,8,9,10,11,12,13,14]. This class of methods solves the transport equation in two steps. The first step tracks back or projects forward one time level along the characteristics of the first order terms of the transport equation and, hence, is Lagrangian in nature. Using the information derived from the first step, in the second step one solves the resultant equations on a fixed, or Eulerian, spatial grid. The Lagrangian step of ELMs eliminates the troublesome first order terms from the algebraic system of equations, and deforming grids associated with purely Lagrangian schemes are avoided.

Recently, Bentley, et. al. [14] introduced the "Eulerian-Lagrangian Least Squares Collocation" method (ELLESCO). In ELLESCO, the time derivative is approximated as a finite

---

*We wish to acknowledge support from the Air Force Office of Scientific Research, Bolling AFB, D.C. and the Ecological Research Division of DOE under Contract No. DE-AC02-83ER60170. We also thank Dr. E. E. Adams of the Ralph M. Parsons Laboratory and Dr. A. M. Baptisa of the Oregon Graduate Center for providing results used in the comparision of Eulerian-Lagrangian methods.
**Formerly at (†), currently at Department of Geology & Geophysics, University of Calgary, Calgary, AB, Canada T2N 1N4.

†College of Engineering and Mathematics, University of Vermont, Burlington, VT, USA 05405.

difference along the characteristics of the first order terms of the equation. The spatial portion of the equation is discretized using cubic Hermite basis functions in conjunction with a least squares collocation method of weighted residuals. Comparison of the results of ELLESCO one-dimensional test problems to the results of other ELMs and the Galerkin finite element method demonstrated the enhanced accuracy of the method. In addition, a Fourier analysis of the procedure was performed [15], and it was shown that ELLESCO demonstrates excellent phase and amplitude behaviour in advective-dominated transport problems.

This paper extends the theory to multi-spatial dimensions. Practical details about the number and location of collocation points, weighting of residuals and implementation of boundary conditions are discussed. ELLESCO performance is compared to the performance of two other ELMs [9,10]. Finally, ELLESCO is used to solve a two-well tracer test problem in order to demonstrate the method's performance in a realistic groundwater engineering application.

2. DEVELOPMENT. 2.1 Transport Equation. Bentley, et. al. [14] developed the ELLESCO equations for the generalized second order parabolic partial differential equation with one spatial dimension. This paper will extend the method to multi-spatial dimensions, but the discussion will be limited to a simplified version of the transport equation often used in water resources modelling. The transport equation for a conservative tracer can be written:

$$\frac{\partial C(\underline{x},t)}{\partial t}+V_i\frac{\partial C(\underline{x},t)}{\partial x_i}-\frac{\partial}{\partial x_i}\left[D_{ij}\frac{\partial C(\underline{x},t)}{\partial x_j}\right]=0 , \tag{1}$$

where repeated indices imply summation, $\underline{x}$ is a spatial coordinate, $C(\underline{x},t)$ is concentration, V_i is an element of the velocity vector and D_{ij} is an element of the dispersion tensor. Associated boundary and initial conditions are:

$$\mathbf{B}C(\underline{x},t) = \overline{b} \qquad \text{on } \delta\Omega , \tag{2}$$

$$C(\underline{x},0) = \overline{C}_0 , \tag{3}$$

where $\mathbf{B}$ is a boundary operator, $\delta\Omega$ is the domain boundary, $\overline{b}$ is the value of the specified boundary condition and $\overline{C}_0$ is the initial concentration field.

When advection dominates, it is well known that classical methods can be troubled by phase errors due to the asymmetry of the first order derivative terms of equation (1) [1]. Transforming to a Lagrangian coordinate system will eliminate the first order spatial derivatives, providing the motivation for Eulerian-Lagrangian methods (ELMs). Accordingly, equation (1) is rewritten in its Lagrangian form:

$$\frac{DC(\underline{x},t)}{Dt}-\frac{\partial}{\partial x_i}\left[D_{ij}\frac{\partial C(\underline{x},t)}{\partial x_j}\right]=0 , \tag{4}$$

where $\dfrac{DC(\underline{x},t)}{Dt}$ is the substantial derivative. The substantial derivative represents the rate of change of the concentration of the fluid as one travels along with the fluid. Equation (4) will be approximated numerically.

2.2 Discretization. The substantial derivative can be approximated at a point, x_k, as the difference between the concentration of the fluid at the current time level and position, (x_k, t_n), and the concentration at the previous time level and the position from which the fluid came in the previous time level, (x_k^*, t_{n-1}), (see Figure 1).

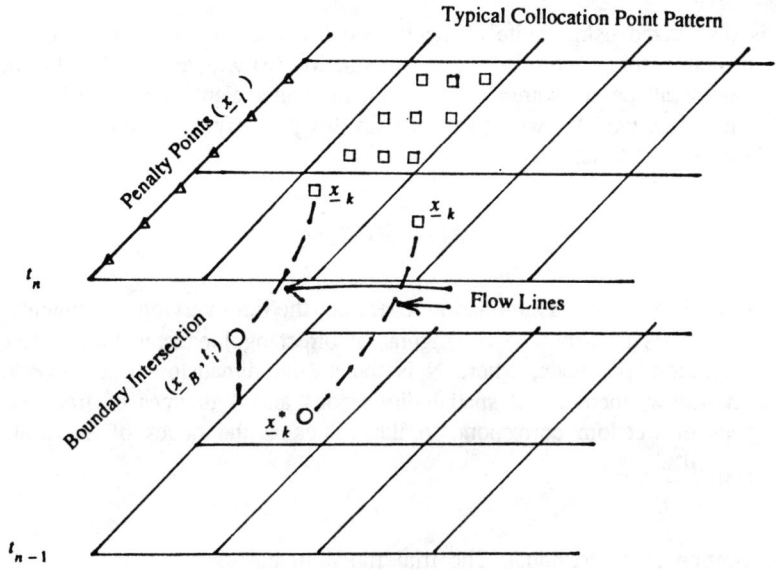

Figure 1. ELLESCO Discretization Geometry. Collocation points from the time level in which the equation is being solved (squares) are tracked back along flow lines (dashed lines) until they intersect the last time level or domain boundary (circles). Penalty points (triangles) can be used to enforce boundary conditions.

In the ELLESCO method, the position in the current time level is given and the associated position in the previous time level is computed. The back-tracked position can be computed by integrating along the velocity field:

$$x_k^* = x_k + \int_{t_n}^{t_{n-1}} V(x(t)) \, dt. \tag{5}$$

The solution of equation (5) is the Lagrangian step of the ELLESCO procedure. An accurate velocity field is required to compute an accurate back-projected position. The creation of this velocity field may be difficult, and many complex issues regarding the generation and use of the velocity field can arise. Within this paper, these issues will be ignored, and the velocity field will be considered exact and known.

The time discretized form of equation (4) for the point located at x_k is:

$$\frac{C(x_k, t_n) - C(x_k^*, t_{n-1})}{\Delta t} - \Theta \frac{\partial}{\partial x_i} \left[D_{ij} \frac{\partial C(x_k, t_n)}{\partial x_j} \right]$$

$$-(1-\Theta)\frac{\partial}{\partial x_i}\left[D_{ij}\frac{\partial C(\underline{x}_k^*, t_{n-1})}{\partial x_j}\right] = 0 \, , \tag{6}$$

where Θ is a time weight and $\Delta t = t_n - t_{n-1}$. Equation (6) is valid for those points that do not track back to the domain boundary. Points that do track to the domain boundary will be discussed later.

Space is discretized using finite elements. Since a least squares procedure is used as a minimization principle, the second derivatives in equation (6) will require that the trial function be C^1 continuous at all points within the domain, including element boundaries. Cubic Hermite basis functions are well known functions with this property [1.]. The concentration field is approximated by the trial function:

$$\hat{C}(\underline{x}, t) = \alpha_J(t)\Phi_J(\underline{x}), \tag{7}$$

where the repeated index implies summation, $\alpha_J(t)$ are the time varying coefficients and $\Phi_J(\underline{x})$ are cubic Hermite basis functions. If rectangular or birectangular elements are used, there are 2^N degrees of freedom per node, where N is the spatial dimension of the domain. In the examples which follow, there are 2 spatial dimensions and 4 degrees of freedom per node. The four degrees of freedom correspond to the values at the nodes of the trial function's $\hat{C}, \dfrac{\partial \hat{C}}{\partial x}, \dfrac{\partial \hat{C}}{\partial y}$ and $\dfrac{\partial^2 \hat{C}}{\partial x \partial y}$.

2.3 Collocation Point Residual. The trial function, equation (7), is substituted into the time discretized transport equation, (6), and, after some rearrangement, the residual associated with a collocation point located at $\underline{x}_k$ can be written:

$$R_k = \alpha_J(t_n)\left[\Phi_J(\underline{x}_k) - \Theta\Delta t \frac{\partial}{\partial x_i}\left[D_{ij}\frac{\partial \Phi_J(\underline{x}_k)}{\partial x_j}\right]\right] -$$

$$\hat{C}(\underline{x}_k^*, t_{n-1}) - (1-\Theta)\Delta t \frac{\partial}{\partial x_i}\left[D_{ij}\frac{\partial \hat{C}(\underline{x}_k^*, t_{n-1})}{\partial x_j}\right]. \tag{8}$$

Minimizing the sum of the squares of the collocation point residuals and the penalty point residuals that will be discussed later forms the basis of the ELLESCO method. The unknowns in equation (8) are the $\alpha_J(t_n)$ coefficients. Referring to Figure 1, the values of the basis functions and their derivatives can be determined once the collocation point location, $\underline{x}_k$, is specified. The back tracked location, $\underline{x}_k^*$, is computed by integrating along the flow field as shown in equation (5). When the location $\underline{x}_k^*$ is known, the value of the trial function, $\hat{C}(\underline{x}_k^*, t_{n-1})$, and its spatial derivatives can be computed from the solution of the previous time step.

A complication arises when the backward track of a collocation point intersects a domain boundary. Referring to Figure 1, one of the collocation point tracks intersects the domain boundary at position X_B and time t_i. Since the time t_i does not correspond to a solution time, a trial function does not exist, and the concentration and its spatial derivatives cannot be evaluated. Bentley, et. al. [14.] detailed a method for handling boundary intersections. Subsequent work has shown that, when advection dominates, this procedure is unneccessarily complicated, and a simpler procedure can be used effectively.

When the domain boundary is intersected, the implicit form, that is $\Theta = 1$, of equation (8) is used, and only the value of the concentration at the back-projected point need be evaluated. The value of the boundary intersection concentration, $C(x_B, t_i)$, is approximated by $\hat{C}(x_B, t_{n-1})$. The approximation works well, because, for most boundaries where fluid enters the domain, the concentration is specified. The time increment is approximated by either $\Delta t = t_n - t_{n-1}$ or, alternatively, by $\Delta t = t_n - t_i$, which scales the amount of dispersion to the length of time the fluid has been in the domain. Often, the form of Δt and the use of the implicit formulation for boundary intersected tracks have little influence on the solution, because both the time weight, Θ, and the time step magnitude, Δt, only affect the dispersive terms of the residual. Since any front that develops at an inflow will be advected away, the concentration field near the inflow will rapidly become nearly uniform, and the dispersion terms of the residuals of collocation points near the inflow become small after the initial time steps. However, if the applied concentration varies temporaly, more care may be necessary.

Any type of boundary can be intersected by the back-track of a collocation point. The approximation to $\hat{C}(x_B, t_i)$ is used only for the calculation of collocation point residuals and is not used to enforce the boundary conditions. Boundary conditions of any type are enforced separately in the matrix equations or with penalty points as discussed in section 2.7.

2.4 Penalty Points.

In addition to the residuals that are defined at regular collocation points, residuals also will be defined at special points referred to as penalty points. Penalty points can be used to enforce boundary conditions or other special conditions. A residual at a penalty point can be written:

$$R_l = \bar{a}_l - \alpha_J(t_n) G \Phi_J(x_l) , \qquad (9)$$

where $\bar{a}_l$ is the specified value of the condition and G is the differential operator representing the condition. For example, if a concentration value were to be specified at x_l, $\bar{a}_l$ would be the value of the concentration that is to be specified and G would be one. The enforcement of boundary conditions is discussed in detail in section 2.7.

2.5 Discrete Equations.

The weighted sum of the squares of the residuals is formed by squaring each collocation point residual, equation (8), and each penalty point residual, equation (9), multiplying each squared value by a weight and summing:

$$\varepsilon = W_K R_k R_k + W_L R_l R_l, \qquad (10)$$

where Ws are weights and repeated indices imply summation.

In order to minimize the sum of the squares of the residuals, the partial derivative of ε with respect to each of the unknown coefficients, $\alpha_I(t_n)$, is set equal to zero:

$$\frac{1}{2} \frac{\partial \varepsilon}{\partial \alpha_I} = W_K R_k \frac{\partial R_k}{\partial \alpha_I} + W_L R_l \frac{\partial R_l}{\partial \alpha_I} = 0 \qquad I = 1, M , \qquad (11)$$

where M is the total number of unknowns in the spatial discretization and α_I corresponds to $\alpha_I(t_n)$. Equation (11) is a system of M algebraic equations and M unknowns.

The system of equations can also be written in matrix form:

$$(\underline{A} + \underline{B}) \underline{\alpha} = \underline{c} + \underline{d} , \qquad (12)$$

where:

$$a_{IJ} = W_K \left[1 - \Theta \Delta t \frac{\partial}{\partial x_i} \left[D_{ij} \frac{\partial}{\partial x_j} \right] \right] \Phi_I(x_k)$$

$$\left[1-\Theta\Delta t\frac{\partial}{\partial x_p}\left[D_{pq}\frac{\partial}{\partial x_q}\right]\right]\Phi_J(\underline{x}_k) \ , \tag{12a}$$

$$b_{IJ} = W_L\,G\Phi_I(\underline{x}_l)\,G\Phi_J(\underline{x}_l) \ , \tag{12b}$$

$$c_I = W_K\left[\hat{C}(\underline{x}_k^*,t_{n-1})+(1-\Theta)\Delta t\frac{\partial}{\partial x_p}\left[D_{pq}\frac{\partial\hat{C}(\underline{x}_k^*,t_{n-1})}{\partial x_q}\right]\right]$$

$$\left[1-\Theta\Delta t\frac{\partial}{\partial x_i}\left[D_{ij}\frac{\partial}{\partial x_j}\right]\right]\Phi_I(\underline{x}_k) \ , \tag{12c}$$

$$d_I = W_L\,\bar{a}_l\,G\Phi_I(\underline{x}_l) \ . \tag{12d}$$

The system matrix of equation (12) is symmetric positive-definite. Before proceeding, it is worth examining the form of equations (10), (11) and (12). The contributions of each type of residual are separated and summed into the system matrix and load vector independently. If a new type of residual is required, it can be formulated and added into the system matrix without disturbing the previously formulated part of the algorithm. In practice, the penalty residuals are a series of different types of residuals corresponding, for example, to the three different types of boundary conditions. The separation of the residual contributions, which is common to all least squares procedures, provides great flexibility in programming.

2.6 Initial Conditions. Two methods are used to specify initial conditions. The first method requires the specification of all of the degrees of freedom within the domain. Thus, the concentration and the appropriate spatial derivatives of the concentration are required at each node. This method is only applied to homogeneous initial conditions or very simple concentration fields.

The second method is to assign initial concentration values to each of the collocation points. The trial function is least squares fitted to the collocation point values. If the initial concentration field has detailed structure, the second method is simpler and more accurate than the first method.

2.7 Boundary Conditions. Boundary conditions can be imposed by one of two methods. Along portions of the domain boundary that follow element boundaries, Dirichlet and Neuman boundary conditions can be imposed directly in the matrix equations. Since the C^1 continuous cubic Hermite basis functions are less familiar than the C^0 continuous Lagrangian basis functions, the procedure for imposing conditions directly in the matrix equations will be described briefly. For more detailed discussions, the reader is referred to Laible and Pinder [16]. Each node in the finite element mesh has 2^N degrees of freedom associated with it, where N is the spatial dimension of the domain. To simplify the discussion, the 2-dimensional case will be discussed with the understanding that the results can be extended to 3 dimensions. Cubic Hermite basis functions are constructed so that the four degrees of freedom associated with each node correspond to the values of $\hat{C}$, $\dfrac{\partial\hat{C}}{\partial x}$, $\dfrac{\partial\hat{C}}{\partial y}$, and $\dfrac{\partial^2\hat{C}}{\partial x\partial y}$ at the node. To set a Dirichlet condition, the value for the degree of freedom associated with the concentration is fixed in the matrix equations. However, there arises an implied condition that should also be set directly in the matrix equations. For example, a constant value of C along a boundary that runs parallel to the x-axis implies that $\dfrac{\partial C}{\partial x} = 0$, and the degree of freedom associated with this derivative should also be fixed in the matrix equations. Similarly, Neumann conditions can be imposed by fixing the value of the degree of freedom associated with the appropriate derivative and imposing an implied cross derivative condition.

The penalty method can be used to impose any type of condition. The penalty point operator, $\mathbf{G}$, of equation (9) becomes the boundary operator, $\mathbf{B}$, of equation (2), and the specified value, $\bar{a}_l$, becomes the specified value of the boundary condition, b, evaluated at the boundary location x_l. The residual associated with the boundary condition is added to the matrix equation as shown in equation (12). As an example, a zero dispersive flux condition is applied along the penalty points in Figure 1. The operator $\mathbf{G}$ becomes $\dfrac{\partial}{\partial x}$ and the value of $\bar{a}_l$ is zero. If the spatial density of the boundary penalty points is great enough, implied conditions are not required. However, if fewer boundary points are used, the implied conditions should also be prescribed at the boundary points. For this example, the implied boundary operator $\mathbf{G}$ is $\dfrac{\partial^2}{\partial x \partial y}$, and $\bar{a}_l$ is again zero. If the implied conditions are enforced, three penalty points per element containing a domain boundary is sufficient. If the implied conditions are not enforced six penalty points per boundary element should be used.

The penalty method of imposing boundary conditions has two advantages compared to the direct setting of degrees of freedom. First, third type boundary conditions can be easily enforced. Second, irregular domain boundaries can be represented on regular, rectangular meshes. Thus, least squares methods can take advantage of the simplicity of regular meshes, without being constrained to regular domain boundaries[16].

2.8 Location And Number Of Collocation Points. The most accurate solutions for a given spatial density of collocation points are obtained when the collocation points are aligned with the lines of flow. However, the improvement in accuracy is accompanied by computational burdens. In all but the simplest cases, the collocation points must be located element by element, depending on the velocity field that is encountered. In transient flow fields, the positions must also be updated as the flow field varies. In contrast, a rectangular pattern, in local element coordinates, is the same for all elements and all time steps.

A related burden arises during the construction of the matrices and vectors of equation (12). Typically, basis functions are defined in terms of elemental or local coordinates[1]. The normal procedure is to compute the contribution to equation (12) from each element and then distribute its contribution to the appropriate locations in the global matrix. If the local coordinates of the collocation point pattern are the same for all elements, the values of the basis functions will be the same for the corresponding points in each element. The same is true for the spatial derivatives with respect to the local coordinates. At the beginning of the program, one set of values and local coordinate derivatives is computed for each collocation point of the pattern. For the nine point example shown in Figure 1, nine sets of sixteen basis function values and nine sets of sixteen each of the required derivatives are computed and saved. The subsequent sets can then be applied to all elements during the matrix construction, enhancing the efficiency of the algorithm. If collocation points follow flow lines, the local coordinates of the collocation points will, in general, vary from element to element and time step to time step, requiring the computation of the values of the basis functions and their derivatives at every collocation point of every time step. These computations increase the effort required to assemble the matrices.

In summary, positioning collocation points in patterns that are similar to the flow stream-lines will improve the accuracy of the computed solution. However, if the local pattern must be varied in order to follow the streamlines, the enhanced accuracy is not generally worth the loss of program simplicity and computational efficiencies that arise from using a single colloca-tion point pattern in all elements. The notable exception occurs in areas near velocity singulari-ties. For example, experience has shown that solutions can be significantly improved by using a radial pattern of collocation points in the immediate vicinity of a pumping well. Figure 9 shows an example of a radial collocation point pattern.

A minimum of four collocation points per element is required or the system will be underdetermined. Consequently, square collocation point patterns of four, nine or sixteen points that are evenly spaced in local coordinates have been used. For an example pattern see Figure 1. It is worth noting that locating collocation points at Gauss points will not necessarily give the optimal accuracy, because the back tracked location of the collocation points will no longer be at the Gauss points. Increasing the number of collocation points per element increases the accuracy of the solution. The largest gain in accuracy is made by increasing the number of points from four to nine followed by increases from nine to sixteen. Increases past sixteen collocation points per element yield only marginal improvement.

Increases in accuracy are accompanied by increased computational requirements. Each collocation point must be tracked back to the last time level, and the tracking effort will be proportional to the number of collocation points. In steady state or simple flow fields, tracking is often a small percentage of the overall computational effort. In complex flow fields, the added tracking effort may be a significant burden. Similarly, the number of computations required for the construction of the $\underline{A}$ matrix and the $\underline{c}$ vector of equation (12) are directly pro-portional to the number of collocation points. In contrast, the bandwidth, rank, and effort required to invert the system matrix are not affected by the number of collocation points.

The significance of the additional computational effort required to increase the number of collocation points is problem dependent. However, the least squares method allows the local insertion of special collocation point patterns in difficult areas with large concentration gra-dients or unusual flow field characteristics. Hence, local accuracy can be enhanced without significant increases in computational burdens.

2.9 Collocation Point Weights. The most accurate results of the ELLESCO method have been obtained by setting the individual collocation point weights, W_K, approximately propor-tional to the size of the area that they represent. The weights for the collocation points within an element are computed by dividing the area of the element by the number of collocation points contained in the element.

Boundary penalty points are weighted proportional to the length of the boundary segment that they represent. In addition, the weights are multiplied by a penalty weight that is set according to the type of boundary condition that is being applied. For example, the penalty points in Figure 1 would be weighted by, $W_L = Wb_L \frac{dyn}{3}$, where dyn is the length of the y-dimension of the element. A typical value for Wb_L is 10^4, and Wb_L should be chosen so that W_L is greater than W_k. As long as W_L is large enough to enforce the boundary condition, ELLESCO results are insensitive to its value, and Wb_L can take a large range of values without affecting the results.

3. COMPARISON OF EULERIAN-LAGRANGIAN METHODS. 3.1 Convection-Diffusion Forum. Results of Problem Set 2 and Problem Set 4 of the Convection-Diffusion Forum (CD Forum) of the VII International Conference on Computational Methods in Water Resources, June, 1988 follow [17]. These problems were compiled in order to provide a

common reference to demonstrate the performance of numerical methods. Problem Set 2 involves the convection of a rotating plume, and the details are found in Appendix I. Problem Set 4 involves diffusion in plane shear flow, and the details are found in Appendix II. The error measures that are used to measure solution accuracy are found in Appendix III.

3.2 Description of Methods. The ELLESCO results are compared to the results of two ELMs that were presented at the CD Forum. The results computed by the ARGOS 2-D MODELLING SYSTEM were presented by J.M. Usseglio- Polatera. The theoretical basis of the program can be found in Holly and Polatera [10]. ARGOS 2D is an industrial computer code used for contaminant transport studies. Like ELLESCO, ARGOS 2-D uses a cubic Hermite basis. However, the Lagrangian calculation is performed by tracking back from nodal locations. Four degrees of freedom are associated with each node, $C, \dfrac{\partial C}{\partial x}, \dfrac{\partial C}{\partial y}$ and $\dfrac{\partial^2 C}{\partial x \partial y}$. Each of these degrees of freedom has a different characteristic path. The characteristics $C, \dfrac{\partial C}{\partial x}$ and $\dfrac{\partial C}{\partial y}$ are tracked from the node to the intersection of the last time level, and their values are computed from the solution of the last time level. The cross derivative is estimated by a divided difference. When the values of the four degrees of freedom are established for all of the nodes, an intermediate solution is created to represent the effects of pure advection. A second step then solves a diffusion problem using the intermediate solution and a Crank-Nicolson scheme. The diffusion calculation uses finite differences combined with an operator splitting technique that produces six sets of tridiagonal matrices. Two sets of ARGOS 2-D results were presented at the CD Forum, the "general procedure" and the "small cloud procedure." The best results of the two sets were obtained using the "small cloud procedure." Since the "small cloud procedure" is, in essence, an adaptive refinement technique, it is not appropriate to compare it to fixed grid methods. Hence, the comparisons presented in this paper will be to the ARGOS 2-D "general procedure" results.

ELLESCO results are also compared to an ELM which will be refered to as ELMLQ. ELMLQ is a method that has been used to study contaminant transport in estuaries and bays. It has been describe by Baptista, et. al. [9] and analysed in detail by Baptista [12]. ELMLQ uses a Lagrangian quadratic basis to discretize space. The location of each node is tracked along the flow field to the previous time level. The values of the trial function at the back tracked locations are assigned to the nodes, forming an intermediate solution. A diffusion problem is then solved using the intermediate solution and an implicit time discretization. It is worth noting that Baptista also presented results from higher order interpolating functions that have, for some problems, given superior results to those of ELMLQ. However, the applicability to irregular domains and meshes is in question, so the more generally applicable ELMLQ results are used for comparison in this paper.

The implicit form of ELLESCO with a rectangular pattern of nine collocation points per element, ELLESCO (3X3) ($\Theta = 1.0$), is used as the main basis of comparison. For selected problems, the results from the Crank-Nicolson form of the nine point rectangular pattern, ELLESCO (3X3) ($\Theta = 0.5$) and the implicit four point rectangular pattern, ELLESCO (2X2) ($\Theta = 1.0$), are also presented. A least squares fit of the trial function to the values of the initial concentration at the collocation points was used as the initial condition. Boundary conditions were imposed by setting the normal derivatives and the cross derivatives to zero in the matrix equations.

Although the test problems have been designed to provide a common reference, differences in the three methods make a direct comparison of the results somewhat misleading. Comparisons of the accuracy of methods are most appropriate when the level of computational effort required by the methods is equivalent. Computational requirements of the three methods discussed in this paper are different and will vary as the problem setting varies. Table 1 summarizes several important parameters that affect the computational effort required by each method to solve the following two test problems.

Assuming A Square Mesh			
METHOD	NODES	UNKNOWNS	TRACKS
ELLESCO (2X2)	N^2	$4N^2$	$4N^2$
ELLESCO (3X3)	N^2	$4N^2$	$9N^2$
ARGOS 2D	$4N^2$	$12N^2$	$12N^2$
ELMLQ	$4N^2$	$4N^2$	$4N^2$
	Θ	MATRIX INVERSION	
ELLESCO (2X2)	1.0	½ Band Width $4N+8$ or Iterative	
ELLESCO (3X3)	1.0	½ Band Width $4N+8$ or Iterative	
ARGOS 2D	0.5	$12N$ Tridiag. Matrices, Length $2N$	
ELMLQ	1.0	½ Band Width $4N+3$ or Iterative	

Table 1. CD Forum Computational Effort Comparison.

If we imagine a square mesh with $2N$ nodes per side, the ARGOS 2D and ELMLQ methods have $4N^2$ nodes in the mesh. ELLESCO used only every other node in the test problems so that it has N^2 nodes in the mesh. The number of degrees of freedom per node of the methods is four, three and one for ELLESCO, ARGOS 2D and ELMLQ respectively. If the measure of spatial discretization is the number of degrees of freedom per unit area, then ELLESCO and ELMLQ have equivalent spatial discretizations. ARGOS 2D has three times the number of degrees of freedom, which is equivalent to a mesh refinement of $\dfrac{1}{\sqrt{3}}$.

Although simple velocity fields are used in the test problems, engineering problems may have complicated velocity fields, and increasing the number of back-tracks can be a significant computational burden. Neglecting the effects of boundary geometry, the number of tracks required by ARGOS 2D and ELMLQ is approximately equal to the number of degrees of freedom. ELLESCO requires the number of collocation points per element times the number of elements which is approximately the number of collocation points per element times N^2. Examining Table 1, ARGOS 2D requires the greatest tracking effort, with three times the number of tracks as ELMLQ. ELLESCO (2X2) requires the same number of tracks as ELMLQ, and ELLESCO (3X3) requires two and one quarter times the tracking effort of ELMLQ and three quarters that of ARGOS 2D.

The effort required to solve the algebraic system of equations differs between the methods. ELLESCO and ELMLQ require inversion of matrices of equal ranks. Direct matrix solvers are also affected by the bandwidth, and ELMLQ has somewhat smaller bandwidth than ELLESCO. ARGOS 2D has a much larger problem to solve, but the method uses an operator splitting technique which creates a set of tridiagonal matrices which can be solved using very efficient techniques. However, efficient iterative techniques, such as preconditioned conjugate gradient method, can be applied to ELLESCO and ELMLQ.

The three forms of ELLESCO have equal rank and bandwidth, so matrix inversion effort does not vary beween the forms. The tracking effort obviously increases as the number of collocation points per element increases. The Crank-Nicolson ELLESCO and the implicit ELLESCO require the evaluation of the trial function at the foot of each characteristic, however, Crank-Nicolson ELLESCO also requires the evaluation of spatial derivatives of the trial function at the foot of each characteristic. The derivative evaluations increase the effort required to form the load vector.

In summary, as the following test problem results are analysed, the following features should be remembered. ELLESCO (2X2) and ELMLQ are equivalent in terms of spatial discretization, tracking requirements and matrix size. ELLESCO (3X3) requires greater tracking effort, but is otherwise similar in its requirements to ELMLQ. ARGOS 2D uses three times the unknowns, which is equivalent to a spatially refined mesh, and it requires more tracking than the other two methods. ARGOS 2D uses an efficient direct solver, but use of efficient iterative solvers with ELLESCO and ELMLQ could negate this apparent advantage.

3.3 Error Measures. The error measures detailed in Appendix III are used to compare the results of the three methods. Briefly, the error ϕ_e is the integral L_2 norm normalized by the mass, and it represents a global error measure. The error ε_e is the difference between the analytic and computed peak values normalized by the analytic peak value. The value of ε_e is related to amplitude error and a positive value indicates an overdamped solution and a negative value an underdamped solution. The error Ψ is the maximum undershoot normalized by the peak amplitude, and it is indicative of phase errors. The error μ_0 is the ratio of the computed mass to the analytic mass, and values different from one indicate mass balance errors.

3.4 Problem Set 2 Results. Problem Set 2 of the CD Forum is pure convection of two dimensional concentration hills by a flow field of counterclockwise rigid body rotation, and the problems are described in detail in Appendix I. The problems consist of completing one full rotation in a number of time steps, and the analytic solution is simply the initial condition (Figure 2). The set consists of six problems, and a summary of the details of the different problem parameters is found in Table 2.

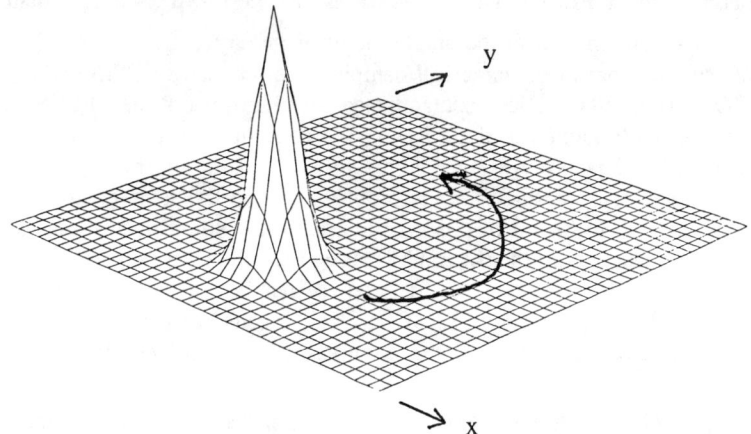

Figure 2. Problem Set 2, CD Forum. A plume is transported one full rotation by advection. ARGOS2D and ELMLQ use the elements shown in the figure. Four squares cover the area of one ELLESCO element.

$\omega = \dfrac{2\pi}{3000}$, Gauss Hill : $\sigma_0 = 264$, Cone Hill : $\tilde{r}_0 = 800$				
PROBLEM	HILL	Δt	NO. STEPS	GRID
2A	GAUSS	100	30	3
2B	CONE	100	30	3
2C	GAUSS	100	30	4
2D	GAUSS	10	300	3
2E	CONE	10	300	3
2F	GAUSS	10	300	4

Table 2. CD Forum Problem Set 2 Parameter Definition.

The Problem 2 error results are presented in Table 3. West to east and south to north concentration profiles for the results of Problems 2A and 2D are presented in Figures 3 and 4 respectively. The ELLESCO results are somewhat damped and out of phase. ELLESCO (3X3) results are superior to the ELLESCO (2X2) results, demonstrating the advantage of adding collocation points. Since there is no diffusion in the problem, the time weight, Θ, is irrelevant.

Fourier analysis [15.] has demonstrated the excellent amplitude preservation characteristics of ELLESCO, but the analysis also indicates that undamped short wavelengths are slightly out of phase. The accumulation of these phase errors is not significant after 30 time steps (Problems 2A-2C and Figure 3). The ELLESCO concentration peaks are well located and have little undershoot, Ψ. After 300 time steps (Problems 2D-2F and Figure 4), the accumulation of phase error in undamped short wavelengths causes trailing oscillations and lagged concentration peaks. Consistent with Fourier analysis, ELLESCO (2X2) is more damped and more severely contaminated by phase errors than ELLESCO (3X3). The effect of irregular node spacing is to increase phase and amplitude errors, but the effects are not severe. In all cases, mass was well conserved by ELLESCO, and the largest mass error, μ_0, was less than 0.2% .

Figures 3 and 4 combined with the amplitude error measure, ε_e, of Table 3 demonstrate that ARGOS 2D suffers from more numerical damping than either ELLESCO (2X2) or (3X3). However, ARGOS 2D results are less contaminated by phase errors than ELLESCO results, and, as demonstrated in Problem 2D, the ARGOS 2D concentration peaks are more precisely located than the ELLESCO peaks. ARGOS 2D results exhibit less undershoot, Ψ, as well. The superior phase propagation of ARGOS 2D is probably due to the numerical damping of the short wavelengths that tend to be out of phase. Both methods exhibit good mass balance. The improved amplitude representation of ELLESCO causes ELLESCO to outperform ARGOS 2D, in the global sense of ϕ_e, for Problems 2A-2C. As the number of time steps increases, as in Problem 2D, ARGOS 2D outperforms ELLESCO (2X2), because of the accumulation of phase error by the latter. ELLESCO (3X3) continues to outperform ARGOS 2D even after many time steps.

The ELMLQ results for Problem Set 2 are much more severely damped than either the ELLESCO (3X3) or (2X2) results. In all error categories, ELLESCO (3X3) performs much better than ELMLQ. ELLESCO (2X2) performs significantly better than ELMLQ for Problems 2A-2C. As noted above, ELLESCO (2X2) has developed serious undershoots and a lagging concentration peak after 300 time steps. ELMLQ's concentration peak is similarly lagged after 300 time steps, but the solution is so seriously damped that the trailing oscillation is smaller

METHOD	ϕ_e	ε_e	ψ	μ_0
Analytic	0	0	0	1
PROBLEM 2A ERRORS				
ELLESCO (2X2)	.4771e-4	.4496e-1	.1683e-2	1.002
ELLESCO (3X3)	.1756e-4	.2144e-1	.3524e-3	1.000
ARGOS 2-D	.5952e-4	.8251e-1	.1155e-2	.9980
ELMLQ	.3575e-3	.4136	.2043e-1	.9853
PROBLEM 2B ERRORS				
ELLESCO (3X3)	.3918e-4	.1237	.1589e-1	.9991
ARGOS 2-D	.5379e-4	.1143	.1089e-1	.9992
ELMLQ	.1611e-3	.2873	.2241e-1	.9883
PROBLEM 2C ERRORS				
ELLESCO (3X3)	.2355e-4	.2463e-1	.8736e-3	1.001
ARGOS 2-D	.7435e-4	.9708e-1	.2014e-2	.9999
ELMLQ	.4022e-3	.4477	.1933e-1	.9998
PROBLEM 2D ERRORS				
ELLESCO (2X2)	.3773e-3	.1688	.1259	.9981
ELLESCO (3X3)	.1722e-3	.9064e-1	.4462e-1	1.000
ARGOS 2-D	.2424e-3	.2975	.1364e-1	.9994
ELMLQ	.7626e-3	.7201	.3474e-1	1.019
PROBLEM 2E ERRORS				
ELLESCO (3X3)	.7017e-4	.1322	.3203e-1	1.000
ARGOS 2-D	Not Available			
ELMLQ	.5156e-3	.6053	.4590e-1	1.024
PROBLEM 2F ERRORS				
ELLESCO (3X3)	.2044e-3	.1289	.5921e-1	1.001
ARGOS 2-D	Not Available			
ELMLQ	.7861e-3	.7490	.2807e-1	1.030

Table 3. Problem Set 2 Error Report.

than that of ELLESCO (2X2) (Figure 4).

As explained in Appendix I, for Problem Set 2 the total accumulated error is due to the interpolation at the base of the characteristics, and it increases with the total number of time steps taken. For Problem Set 2, ELM results generated using fewer large time steps (Figure 3) are superior to the results generated using more small time steps (Figure 4). In contrast, the results generated by classical schemes improve as the time step decreases, because the time truncation error associated with the first order advection term decreases.

3.5 Problem Set 4 Results. Problem Set 4 of the CD Forum is the transport of a plume in shear flow with constant diffusion, and the problems are described in detail in Appendix II. Figure 5 shows a plan view of the problem, and a summary of the details of the individual

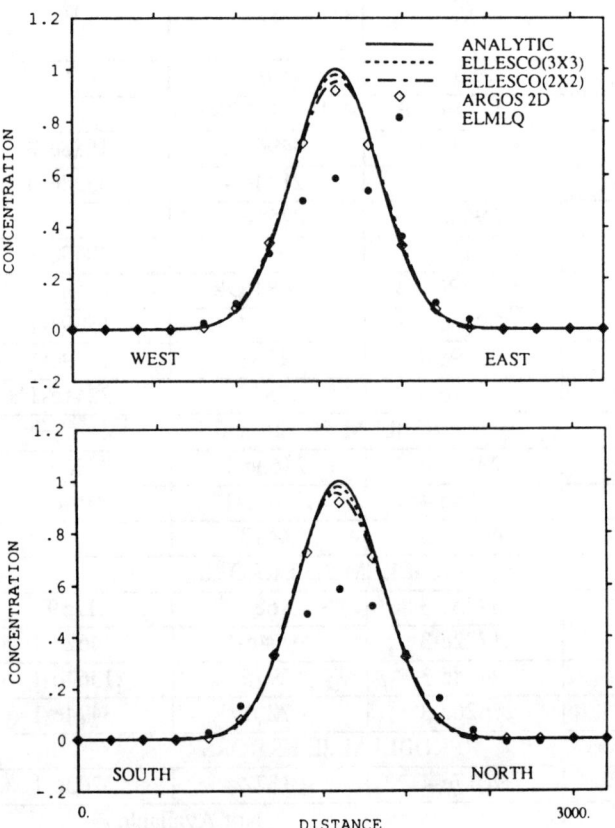

Figure 3. Problem 2A Concentration Profiles. Concentration profiles through the center of the analytic solution (solid lines) are shown as broken lines for ELLESCO results and symbols for ARGOS 2D and ELMLQ node values.

problems is found in Table 4.

The error results for times $t = t_0+2400$ and $t = t_0+7200$ are found in Tables 5A and 5B and 6A and 6B respectively. Plume cross sections of the initial conditions of ELLESCO (3X3) for Problem 4A, which are the same as those for Problems 4G and 4H, are found in Figure 6. Plume cross sections of Problem 4A results are found in Figures 7 and 8 for times t_0+2400 and t_0+7200 respectively.

As demonstrated in Figure 6, ELLESCO (3X3) initial conditions are damped, with sharper peaked initial plumes, such as those of Problems 4C and 4D, more damped than the broader initial plumes. In general, the solutions stay overdamped as time progresses, and, although the magnitude of the peak amplitude error decreases, the percentage amplitude error, ε_e, is little changed. The exception is Problem 4F in which the diffusion, D, was tripled to 30. Fourier analysis results [15] have demonstrated that, when diffusion becomes large and the implicit formulation is used, the shorter spatial wavelengths will be underdamped, explaining the slightly underdamped result of Problem 4F at $t = t_0+7200$.

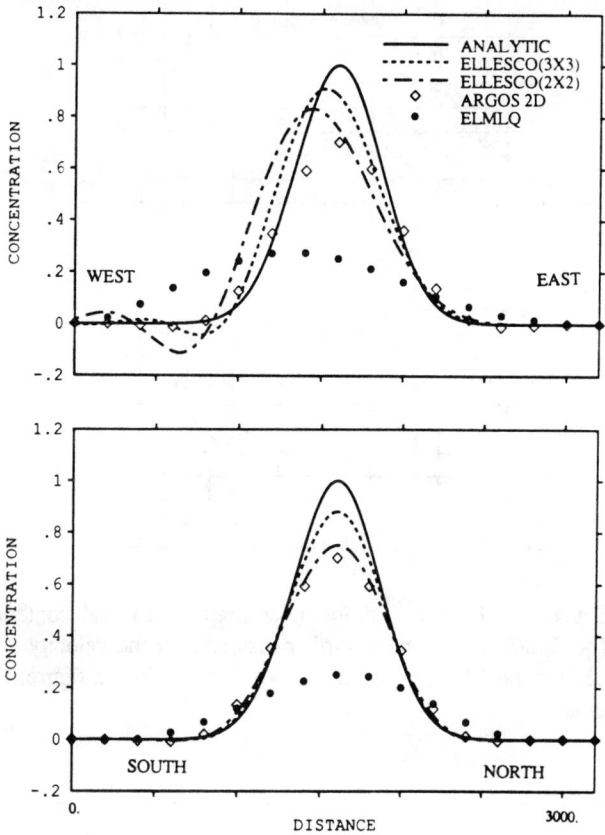

Figure 4. Problem 2D Concentration Profiles. Concentration profiles through the center of the analytic solution (solid lines) are shown as broken lines for ELLESCO results and symbols for ARGOS 2D and ELMLQ node values.

Mass is well conserved in the ELLESCO results, with the largest error of 0.07% being a result of mass error in the initial condition of the sharp peaked Problem 4D. Similarly, the results contain very minor undershoot, Ψ, with a maximum of 0.05% of peak amplitude for Problem 4I which used an irregular grid. The phase errors found in Problem Set 2 are not found in Problem Set 4, because the short wavelength components that are out of phase have been damped by the diffusion.

Global errors, ϕ_e, are small and decrease as time progresses, because amplitudes decrease due to diffusion. Predictably, global errors are larger for narrower plumes, more shear, less diffusion and the irregular grid. As the time step is increased (Problems 4G and 4H), the global error increases while peak amplitude and undershoot errors decrease.

Error results for Problem 4A are presented for ELLESCO (3X3) ($\Theta=1.0$), ELLESCO (3X3) ($\Theta=0.5$) and ELLESCO (2X2) ($\Theta=1.0$). The error measures demonstrate that the ELLESCO (3X3) ($\Theta=1.0$) results are the best of the three. As stated, implicit ELLESCO results are underdamped at short wavelengths, partially compensating for the somewhat dispersed initial condition. The ELLESCO (3X3) ($\Theta=0.5$) does not underdamp the short

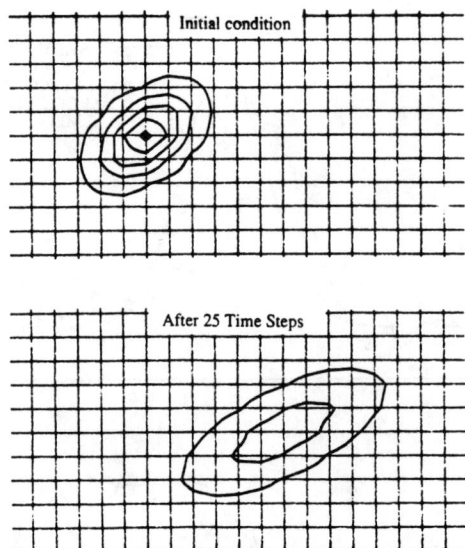

Figure 5. Problem Set 4, CD Forum. Minimum contour is 0.1 and contour interval is 0.2. After 25 time steps (t_0+2400), the plume is further distorted by the velocity shear and flattened by diffusion. ARGOS 2D and ELMLQ use the node locations in the figure. Four squares cover one ELLESCO element.

			$V_0 = 0.5$			
PROBLEM	λ	Δt	t_0	D	GRID	COMMENT
4A	5.0e-4	96	2400	10	7	
4B	0.0	96	2400	10	7	NO SHEAR
4C	1.0e-3	96	2400	10	7	MORE SHEAR
4D	5.0e-4	96	1440	10	7	NARROWER PLUME
4E	5.0e-4	96	4800	10	7	BROADER PLUME
4F	5.0e-4	96	2400	30	7	MORE DIFFUSION
4G	5.0e-4	240	2400	10	7	LARGER Δt
4H	5.0e-4	480	2400	10	7	STILL LARGER Δt
4I	5.0e-4	96	2400	10	8	IRREGULAR GRID

Table 4. CD Forum Problem Set 4 Parameter Definition.

wavelengths. Without an error to compensate for the dispersed initial condition, the Crank-Nicolson ELLESCO results are more dispersed than the implicit results, explaining the superior implicit error results. ELLESCO (3X3) ($\Theta=0.5$) also exhibits slightly larger undershoot at

Method	ϕ_e	ε_e	ψ	μ_0
Analytic	0	0	0	1
PROBLEM 4A ($T = T_0 + 2400$)				
ELLESCO (3X3) ($\Theta = 1.0$)	.1230e-4	.2506e-1	.4646e-4	1.001
ELLESCO (3X3) ($\Theta = 0.5$)	.1362e-4	.3196e-1	.1662e-3	1.000
ELLESCO (2X2) ($\Theta = 1.0$)	.2429e-4	.3141e-1	.2674e-4	1.000
ARGOS 2-D	.3305e-4	-.5372e-1	.8839e-6	1.000
ELMLQ	.8814e-4	.1247	.1923e-1	1.002
PROBLEM 4B ($T = T_0 + 2400$), NO SHEAR				
ELLESCO (3X3) ($\Theta = 1.0$)	.1035e-4	.2027e-2	.4565e-4	1.000
ARGOS 2-D	.2150e-4	-.2925e-1	.4797e-4	1.000
ELMLQ	.1623e-3	.1890	.2986e-1	.9987
PROBLEM 4C ($T = T_0 + 2400$), MORE SHEAR				
ELLESCO (3X3) ($\Theta = 1.0$)	.2246e-4	.5803e-1	.6188e-3	1.000
ARGOS 2-D	.5336e-4	-.1206	.1516e-11	1.000
ELMLQ	.5519e-4	.6873e-1	.1326e-1	1.002
PROBLEM 4D ($T = T_0 + 2400$), NARROWER PLUME				
ELLESCO (3X3) ($\Theta = 1.0$)	.1711e-4	.2843e-1	.2808e-3	.9993
ARGOS 2-D	.4042e-4	-.7435e-1	.1991e-5	1.000
ELMLQ	.6923e-3	.3707	.1638e-1	1.015
PROBLEM 4E ($T = T_0 + 2400$), BROADER PLUME				
ELLESCO (3X3) ($\Theta = 1.0$)	.7528e-5	.2032e-1	.3479e-4	1.000
ARGOS 2-D	.1906e-4	-.3864e-1	.4478e-12	1.000
ELMLQ	Not Available			

Table 5A. Problem Set 4, $t = t_0+2400$ Error Report.

$t = t_0+2400$, consistent with the poor phase behaviour of very short wavelengths predicted by the Fourier analysis of Crank-Nicolson ELLESCO. Not surprisingly, the nine collocation point pattern results are generally superior to the ELLESCO (2X2) ($\Theta=1.0$) results.

For comparison, ARGOS 2D and ELMLQ results are also shown in Figures 7 and 8, and their error results are presented in Tables 5 and 6. Once again, all three methods have good mass balance. All forms of ELLESCO have less global error than ARGOS 2D. ARGOS 2D results are insufficiently diffused, and the subsequent amplitude errors, ε_e, of the method are more severe than those of ELLESCO. Although not visible in the cross sections presented in Figures 7 and 8, ARGOS 2D is less contaminated by undershoot, Ψ, again demonstrating that the ARGOS 2D results have less phase error than those of ELLESCO. ELLESCO results have less error than ELMLQ in all categories except mass balance. Except for Problems 4G and 4H, ELMLQ results are more severely overdamped than ELLESCO results.

ELMs suffer from two numerical errors that are affected by the magnitude of the time step. The first error is an interpolation error which is accumulated with each estimation of the trial function at the back-tracked location. As the time step increases in magnitude, less steps

Method	ϕ_e	ε_e	ψ	μ_0
Analytic	0	0	0	1
PROBLEM 4F (T = T_0 + 2400), MORE DIFFUSION				
ELLESCO (3X3) (Θ = 1.0)	.3510e-6	.5048e-2	.9778e-6	1.000
ARGOS 2-D	.5331e-5	-.1476e-1	.3123e-17	1.000
ELMLQ	Not Available			
PROBLEM 4G (T = T_0 + 2400), LARGER Δt				
ELLESCO (3X3) (Θ = 1.0)	.1322e-4	.1741e-1	.3986e-4	1.000
ARGOS 2-D	.3137e-4	-.5421e-1	.6385e-9	1.000
ELMLQ	.4681e-4	.6175e-1	.8049e-2	1.001
PROBLEM 4H (T = T_0 + 2400), STILL LARGER Δt				
ELLESCO (3X3) (Θ = 1.0)	.1927e-4	.7605e-2	.2803e-4	1.000
ELLESCO (3X3) (Θ = 0.5)	.1534e-4	.3533e-1	.2389e-3	.9999
ARGOS 2-D	.2964e-4	-.4616e-1	.0000	1.000
ELMLQ	.3680e-4	-.2446e-1	.2088e-2	1.003
PROBLEM 4I (T = T_0 + 2400), IRREGULAR GRID				
ELLESCO (3X3) (Θ = 1.0)	.1625e-4	.2013e-1	.5117e-3	1.000
ARGOS 2-D	.3703e-4	-.6096e-1	.1236e-6	1.000
ELMLQ	Not Available			

Table 5B. Problem Set 4, $t = t_0 + 2400$ Error Report.

are required, and the accumulated interpolation error tends to decrease. The second error is the truncation error associated with the time derivative approximation. The truncation error increases as the the time step magnitude increases.

The large increases in time step in Problems 4G and 4H have only minimal effect on the ELLESCO and ARGOS 2D results, reflecting the accurate interpolation functions used by the methods. ELMLQ changes from being overdamped to underdamped in Problem 4H, and, in general, ELMLQ showed significant improvement as the time step increased. Baptista [12.] demonstrated that the ELMLQ had an optimal time step size of $\Delta t = 200$ to 400 for this problem, and that the improved global errors are attributed to the tendency of truncation errors to cancel interpolation errors.

As the time step increases, Crank-Nicolson ELLESCO is expected to exhibit some advantage over implicit ELLESCO. As can be seen from Problem 4H, the global error, ϕ_e, is less for ELLESCO (3X3) (Θ = 0.5) than for ELLESCO (3X3) (Θ = 1.0).

3.6 Comparison Summary. In general, ELLESCO performs somewhat better than ARGOS 2D in Problem Set 2, even though the latter method used three times the number of degrees of freedom. ELLESCO results are much better than ELMLQ results generated with equivalent spatial discretizations and tracking effort. ARGOS 2D exhibits the best phase behaviour of the three methods, and ELLESCO's phase behaviour is superior to ELMLQ phase behaviour. ELLESCO's enhanced performance is due to less numerical damping of the short spatial wavelengths. However, after many time steps, the phase errors of the undamped short wavelengths can lead to trailing oscillations, especially when only four collocation points per

Method	ϕ_e	ε_e	ψ	μ_0
Analytic	0	0	0	1
PROBLEM 4A (T = T_0 + 7200)				
ELLESCO (3X3) (Θ = 1.0)	.7813e-5	.2826e-1	.2616e-4	1.000
ELLESCO (3X3) (Θ = 0.5)	.7830e-5	.3297e-1	.2795e-4	1.000
ELLESCO (2X2) (Θ = 1.0)	.1499e-4	.3546e-1	.9276e-6	1.000
ARGOS 2-D	.2412e-4	-.5763e-1	.1017e-10	1.000
ELMLQ	.2844e-4	.6414e-1	.5485e-2	1.001
PROBLEM 4B (T = T_0 + 7200), NO SHEAR				
ELLESCO (3X3) (Θ = 1.0)	.6269e-5	-.8286e-2	.2240e-6	1.000
ARGOS 2-D	.1142e-4	-.2153e-1	.6572e-5	1.000
ELMLQ	.1101e-3	.1720	.3068e-1	.9987
PROBLEM 4C (T = T_0 + 7200), MORE SHEAR				
ELLESCO (3X3) (Θ = 1.0)	.1193e-4	.4868e-1	.3578e-4	1.000
ARGOS 2-D	.1900e-4	-.9895e-1	.0000	1.000
ELMLQ	.1812e-4	-.1907e-1	.1006e-2	1.002
PROBLEM 4D (T = T_0 + 7200), NARROWER PLUME				
ELLESCO (3X3) (Θ = 1.0)	.8890e-5	.3047e-1	.3172e-4	.9993
ARGOS 2-D	.1944e-4	-.6687e-1	.8566e-18	1.000
ELMLQ	.2294e-3	.2130	.5537e-2	1.015
PROBLEM 4E (T = T_0 + 7200), BROADER PLUME				
ELLESCO (3X3) (Θ = 1.0)	.5326e-5	.2289e-1	.1962e-4	1.000
ARGOS 2-D	.1595e-4	-.4338e-1	.3853e-17	1.000
ELMLQ	Not Available			

Table 6A. Problem Set 4, $t = t_0 + 7200$ Error Report.

element are used.

ELLESCO has somewhat outperformed ARGOS 2D in Problem Set 4. The ARGOS 2D results were inferior because they were insufficiently diffused. Although phase errors were insignificant for all methods in this problem set, ARGOS 2D again exhibited the best phase behaviour. ELLESCO, including ELLESCO (2X2) (Θ = 1.0) , has significantly outperformed ELMLQ, because the ELMLQ results were more severely over damped.

4. TWO-WELL TRACER TEST 4.1 Problem Description. ELLESCO has been shown to perform well in test problems with continuous velocity fields and constant or no dispersion. However, engineering problems typically require the use of dispersion that varies in magnitude spatially and velocity fields with singularities, such as those associated with pumping or injection wells. The two-well tracer test is a classic problem in groundwater transport, and it will be used to demonstrate ELLESCO performance under more realistic conditions.

The following example is based on laboratory tracer experiments by Hoopes and Harleman [18]. An injection and pumping well are separated by 61 cm, and each is operated at a rate

Method	ϕ_e	ε_e	ψ	μ_0
Analytic	0	0	0	1
PROBLEM 4F (T = T$_0$ + 7200), MORE DIFFUSION				
ELLESCO (3X3) (Θ = 1.0)	.1767e-5	-.5215e-2	.4664e-6	1.000
ARGOS 2-D	.2755e-5	-.1396e-1	.0000	1.000
ELMLQ	Not Available			
PROBLEM 4G (T = T$_0$ + 7200), LARGER Δt				
ELLESCO (3X3) (Θ = 1.0)	.9245e-5	.2381e-1	.2049e-4	1.000
ARGOS 2-D	.1593e-4	-.5142e-1	.0000	1.000
ELMLQ	.1457e-4	.1635e-1	.7234e-3	1.000
PROBLEM 4H (T = T$_0$ + 7200), STILL LARGER Δt				
ELLESCO (3X3) (Θ = 1.0)	.1326e-4	.1592e-1	.9128e-5	1.000
ELLESCO (3X3) (Θ = 0.5)	.8347e-5	.3619e-1	.6234e-4	.9999
ARGOS 2-D	.1512e-4	-.4176e-1	.0000	1.000
ELMLQ	.2103e-4	-.5405e-1	.1396e-4	1.005
PROBLEM 4I (T = T$_0$ + 7200), IRREGULAR GRID				
ELLESCO (3X3) (Θ = 1.0)	.1056e-4	-.4058e-1	.1732e-3	1.000
ARGOS 2-D	.1971e-4	-.6305e-1	.2090e-16	1.000
ELMLQ	Not Available			

Table 6B. Problem Set 4, $t = t_0+7200$ Error Report.

of 2.339 cm^3/sec. The aquifer is 8.9 cm thick and has a porosity of 0.374. With the flow field at steady state, the concentration of the injected fluid is set to one at time zero. The main objective of the calculation is to predict the concentration breakthrough curve of the pumping well. The aquifer will be modelled with two spatial dimensions.

Dispersion is modelled with the Scheidegger approximation:

$$D_{xx} = a_l \frac{V_x^2}{V} + a_t \frac{V_y^2}{V} + D_0 \,, \tag{13a}$$

$$D_{yy} = a_t \frac{V_x^2}{V} + a_l \frac{V_y^2}{V} + D_0 \,, \tag{13b}$$

$$D_{xy} = D_{yx} = (a_l - a_t) \frac{V_x V_y}{V} \,, \tag{13c}$$

where V_x and V_y are fluid velocities in the x and y directions respectively, V is the magnitude of the fluid velocity, D_0 is the molecular diffusion constant and a_l and a_t are the longitudinal and transverse dispersivities respectively. Using the Scheidegger approximation, Hoopes and Harleman [18] have derived analytic solutions for the concentration distribution resulting from the two-well tracer test problem.

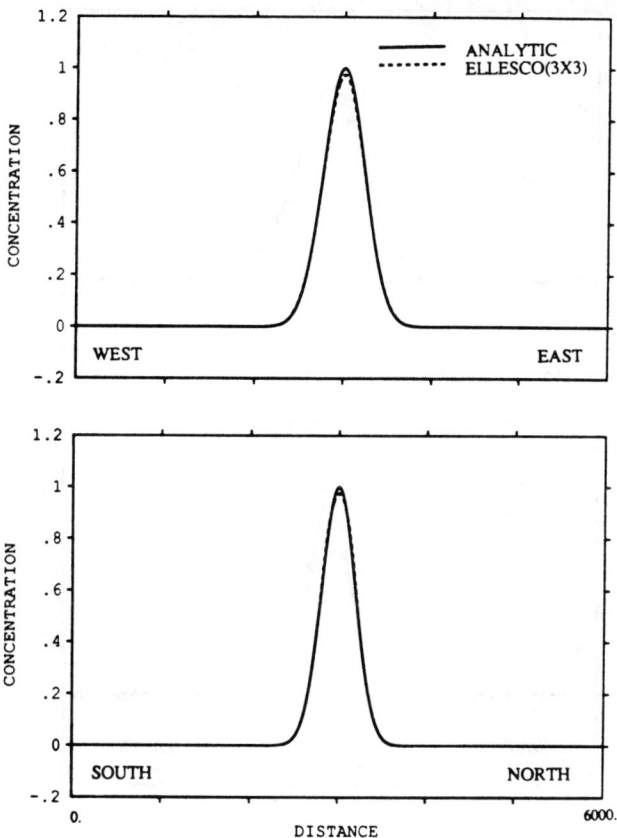

Figure 6. ELLESCO (3X3) Initial Conditions for Problems 4A, 4G and 4H. The ELLESCO initial condition computed by a least squares fit of collocation point values is overdamped.

Huyakorn, et. al. [19.] solved the problem with Galerkin finite elements using the grid shown in Figure 9. The results were contaminated by serious oscillations at a_l = 0.294 cm and milder oscillations at a_l = 2.94 cm. They subsequently improved on those results by using a curvilinear finite element procedure.

4.2 ELLESCO Solution. Using the grid in Figure 9, the implicit form of ELLESCO was used to solve the problem. A rectangular pattern of four collocation points per element was used except in eight elements surrounding the injection well and another eight elements surrounding the pumping well. In the elements surrounding the wells, a radial pattern of collocation points, Figure 9, was found to improve the solution, because aligning collocation points along flow lines and increasing the spatial density of collocation points improves the accuracy of ELLESCO. A zero normal derivative was enforced on all grid boundaries using six penalty points per element side. Boundary residuals were weighted by the length of the boundary segment that they represent multiplied by 10^4. The initial time step interval was Δt = 100sec. The interval was doubled every five time steps until a maximum of Δt = 800sec.

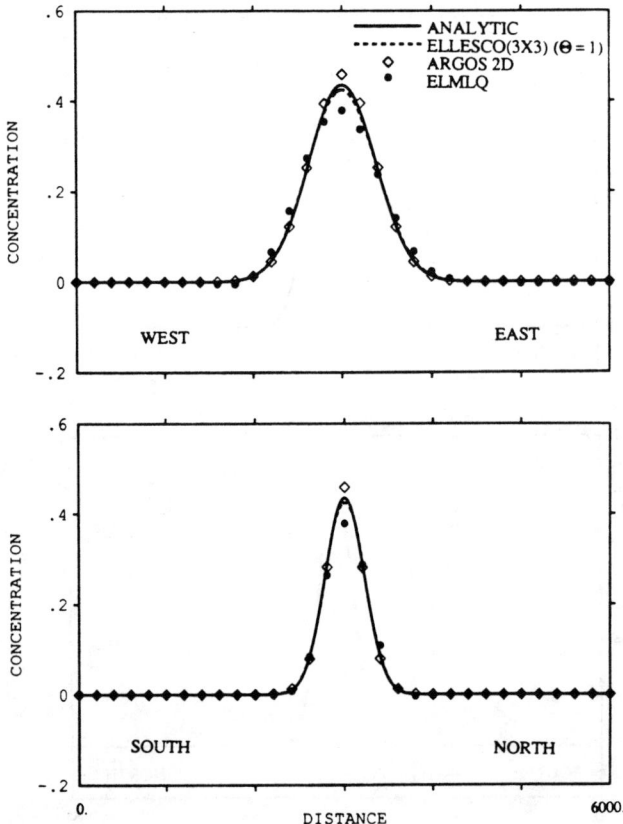

Figure 7. Problem 4A, $t=t_0+2400$ Concentration Profiles. Concentration profiles through the peak of the analytic solution (solid lines) are shown as broken lines for ELLESCO and symbols for ARGOS 2D and ELMLQ node values.

Velocity singularities exist at the wells, and tracking back to the injection well is quite challenging. Consequently, a ring of points surrounding the injection well was tracked forward to delineate a "capture zone." All of the collocation points that fall within the capture zone are assumed to have originated from the injection well, and the concentrations, $\hat{C}(x_k^*, t_{n-1})$, at the foot of the flow path of the captured collocation points are set to one. Theoretically, captured collocation points should use a smaller Δt in equation (8) to reflect the fact that the time in the domain is less that a full time step. In practice, the use of a scaled Δt leads to no significant difference in the solutions, and the results that follow were calculated using the same Δt for all collocation point residuals.

Results for the no dispersion case are presented in Figure 10. In the no dispersion case, ELLESCO reduces to a least squares fit of the advected solution to the collocation points by piece-wise cubic Hermite polynomials. The pumping well concentration breakthrough curve is quite accurate at early time (Figure 10A), but it develops a mild oscillation starting at dimensionless time one. Figure 10B shows concentration profiles from the point midway between the wells through the pumping well for two times. The pumping well causes a step in the

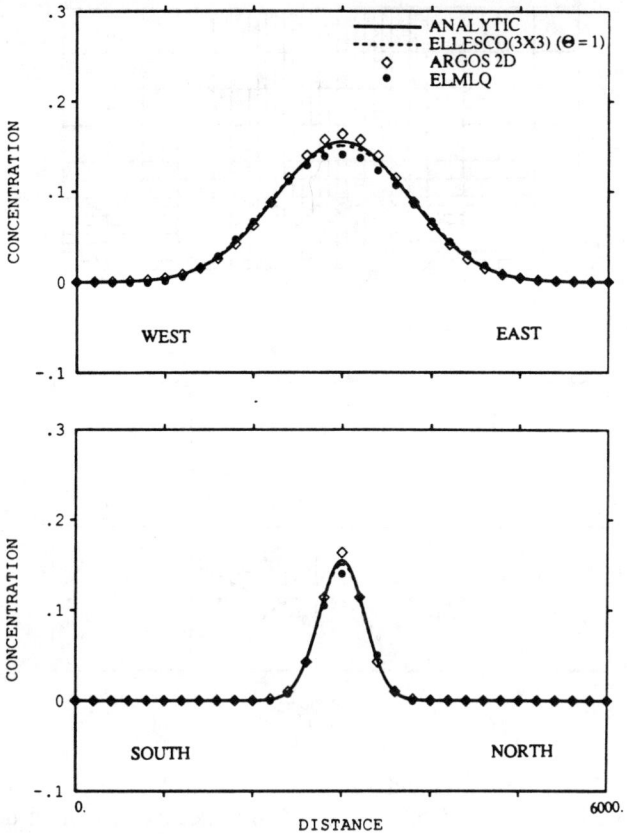

Figure 8. Problem 4A, $t=t_0+7200$ Concentration Profiles. Concentration profiles through the peak of the analytic solution (solid lines) are shown as broken lines for ELLESCO and symbols for ARGOS 2D and ELMLQ node values.

concentration profile at the well, and the solutions around the well are contaminated by severe Gibb's phenomenon oscillations. The severity of the oscillations appears to increase as the tracer from more streamlines is captured by the well.

Figure 10C shows concentration profiles along a line starting at the injection well and striking 45 degrees from the line connecting the two wells. At $t = 8300$sec, the calculated front is accurately located, but has an oscillation. As time progresses, the computed front becomes distorted and the amplitude of the oscillations grow, because the front has passed into the coarser part of the mesh.

The results for the case of $a_l = 0.294$cm, $a_t = 0$, $D_0 = 0$ are shown in Figure 11. The computed pumping well breakthrough curve accurately duplicates the analytic curve (Figure 11A). Concentration profiles through the pumping well location show that the oscillations have almost vanished, and the local concentration gradients are much milder than those of the no dispersion example. At early time, the 45 degree concentration profile is almost an exact duplicate of the analytic. At later times, as the front enters the coarser part of the mesh, oscillations develop, and the front is locally deformed.

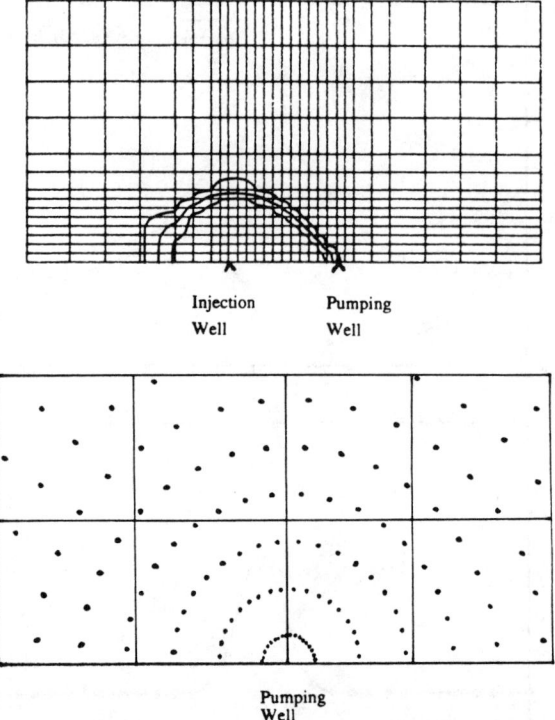

Figure 9. Two Well Tracer Test Grid and Radial Collocation Point Pattern. The injection well and pumping well are located at (-30.5,0) and (30.5,0) respectively. A front develops as solute is introduced at the injection well. The grid is symmetric about x = 0.0cm, and nodes are located at x = 0.0, 5., 10., 15., 20., 25., 30.5, 35., 40., 50., 60., 80., 100., 120., 145. cm, the symmetric negative locations and y = 0.0, 5., 10., 15., 20., 25., 30., 35., 40., 50., 60., 80., 100., 120., 145. cm. A radial collocation point pattern (dots) is used in the eight elements surrounding the pumping well. A similar pattern surrounds the injection well.

The results of the two well tracer test calculations demonstrate the accuracy of ELLESCO in advective dominated flow systems. With purely advective flow, the computed pumping well concentration breakthrough curve exhibits only a mild oscillation. With a more realistic dispersivity, the pumping well breakthrough curve is well predicted. The front is well modelled in the refined portion of the mesh. However, as the front enters the coarser portion of the mesh, it deforms, indicating that an adaptive local grid refinement strategy would be beneficial.

5. CONCLUSIONS. The ELLESCO method has been developed to solve sharp front problems in advective-dominated transport sytems. The asymmetric first order terms associated with advection cause classical methods to suffer from phase and amplitude errors unless very small space and time discretization intervals are used. By using a tracking or Lagrangian step, ELLESCO eliminates the troublesome first order portion of the operator. Consequently ELLESCO can use coarser spatial and temporal discretizations than classical methods.

In addition to enhanced accuracy, ELLESCO has several appealing features. Individual collocation point locations are tracked over single time steps, and continuous particle tracking

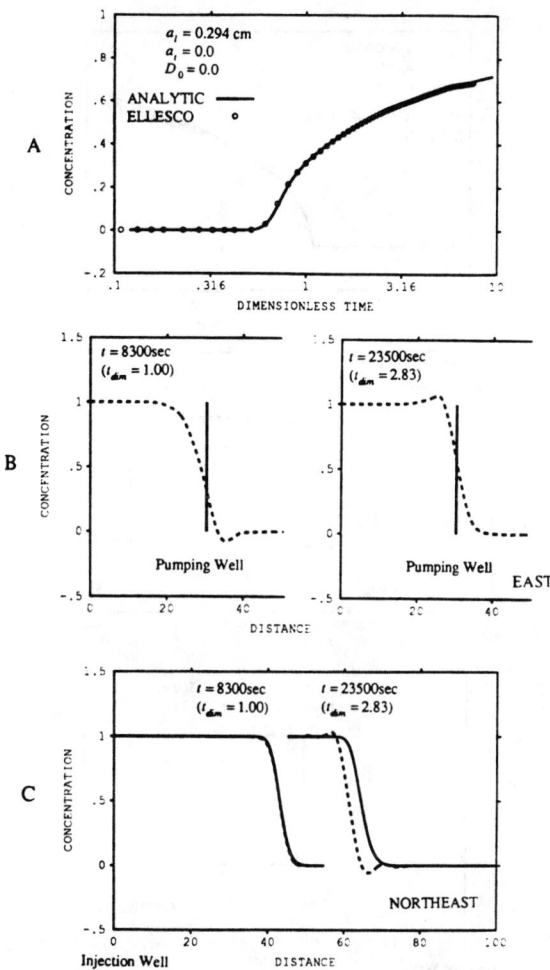

Figure 10. Two Well Tracer Test Results, $a_l = 0.0$. A. Analytic (solid line) and ELLESCO (circles) pumping well concentration breakthrough curves. B. ELLESCO concentration profiles from the center point between the wells through the pumping well. C. Analytic (solid line) and ELLESCO (broken line) concentration profiles along the line starting at the injection well and striking 45 degrees from the line connecting the wells.

is not required. Although Lagrangian in form, ELLESCO does not use a deforming grid. The number and location of the collocation points can be varied locally to enhance accuracy in areas of complex flow, while maintaining simple, computationally efficient collocation point patterns in the remainder of the domain. Finally, by enforcing boundary conditions with penalty points, irregular domain problems can be solved on regular meshes.

ELLESCO has been compared to two ELMs that have been used to solve engineering problems. It was demonstrated that ELLESCO is more accurate than both ELMLQ and ARGOS 2D when equivalent spatial discretizations are used. The ELLESCO results are less artificially diffused because of superior performance of the method in the advective step. ARGOS 2D demonstrates the best phase behaviour of the three methods. Mass was well

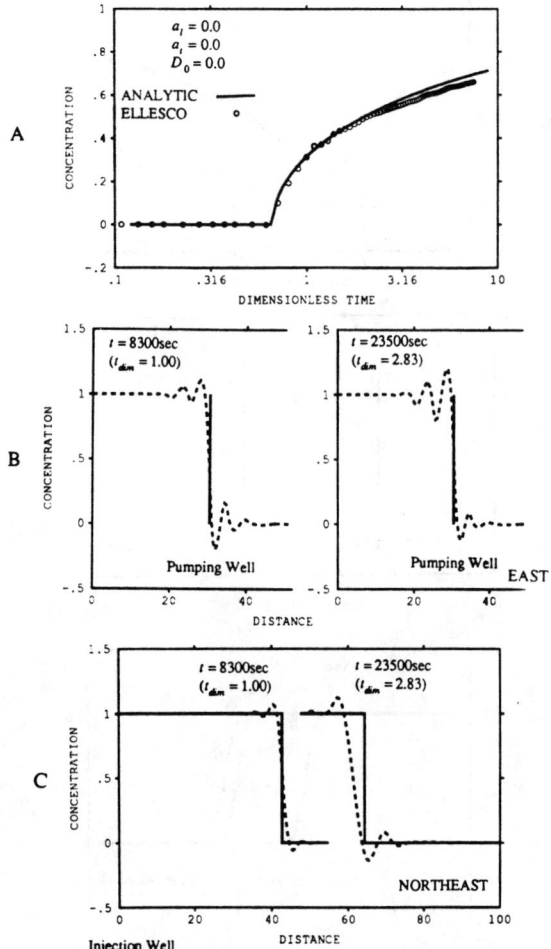

Figure 11. Two Well Tracer Test Results, $a_l = 0.294cm$. A. Analytic (solid line) and EL-LESCO (circles) pumping well concentration breakthrough curves. B. ELLESCO concentration profiles from the center point between the wells through the pumping well. C. Analytic (solid line) and ELLESCO (broken line) concentration profiles along the line starting at the injection well and striking 45 degrees from the line connecting the wells.

conserved by all methods.

When advection dominates, spatial errors due to interpolation at the base of the characteristics will generally be larger than time truncation errors. Consequently, use of the more accurate Crank-Nicolson formulation is typically not worth the extra computational effort, and the implicit form of ELLESCO is the normal method of choice.

Finally, ELLESCO has been shown to perform well with a realistic engineering problem. ELLESCO performs well in the presence of spatially variable physical dispersion, and velocity singularities associated with injection wells are handled well by calculating collocation point capture zones. ELLESCO is characterized by accurate front propagation and concentration breakthrough curves, but performance in the coarser portions of the mesh indicate that local, adaptive grid refinement would be useful.

Appendix I. CD Forum Problem Set 2. The objective of the six problems in this set is to transport by convection concentration hills in a flow field of counterclockwise rigid body rotation. The problems consist of one complete rotation, and the solution is the initial condition. The angular velocity of the rigid body rotation, ω, is $\dfrac{2\pi}{3000}$. Two initial conditions are used. A Gauss hill:

$$C_0(x,y) = \exp\left[-\frac{(x-x_0)^2+(y-y_0)^2}{2\sigma_0^2}\right],\tag{AI.1}$$

where $C_0(x,y)$ is both the initial condition and the solution, (x_0,y_0) is the concentration hill center of mass and σ_0 is the standard deviation of the concentration hill. The second initial condition is a cone hill:

$$C_0(x,y) = 1-\left[\frac{(x-x_0)^2+(y-y_0)^2}{\bar{r}_0^2}\right]^{1/2},\tag{AI.2}$$

for $(x-x_0)^2+(y-y_0)^2 < \bar{r}_0^2$ and zero elsewhere. The parameter $\bar{r}_0$ is the radius of the base of the cone and the other values are as specified above.

Two grids are specified. The first is a regular grid, and, for consistency with the CD Forum nomenclature, is called Grid 3:

$$xn(i,j) = 200(i-1)-3400 \qquad i = 1,35$$

$$yn(i,j) = 200(j-1)-3400 \qquad j = 1,35\,,\tag{AI.3}$$

where $xn(i,j)$, $yn(i,j)$, are the nodal coordinates and i,j are the nodal indices. An irregular grid is also used, and it is called Grid 4:

$$xn(i+1,j)-xn(i,j) = 200-50\cos\left[\frac{\pi(i-1)}{35}\right] \qquad i = 1,34$$

$$yn(i,j+1)-yn(i,j) = 200-50\cos\left[\frac{\pi(j-1)}{35}\right] \qquad j = 1,34\,.\tag{AI.4}$$

The solutions of ARGOS 2D and ELMLQ were computed using the basic node spacing of 200. The node locations used in the ELLESCO computations can be arrived at by replacing the values of 200 in equations (AI.3) and (AI.4) with the value of 400, the values of 50 in equation (AI.4) with 100 and using an index sweep of $i,j = 1,18$ in place of $i,j = 1,35$.

The boundary condition at inflow boundaries is zero concentration, and the boundary condition at outflow boundaries is left to the individual modeller to specify in a consistent manner.

The details of the six problems are found in Table 2. Solutions of classical methods such as Galerkin finite elements or finite differences which are generated using the current order of spatial discretization are contaminated by serious phase errors which cause large oscillations [Glass and Rodi 1982]. In addition to being a difficult problem for classical methods, the plume crosses elements at all angles, and the results will demonstrate problems due to grid orientation. Problems 2C and 2F also test for accuracy problems arising from non-uniform nodal spacings. Finally, problems 2D through 2F have small time increments and many time steps. Classical methods tend to improve with smaller time steps, because time truncation errors are reduced. Contrarily, for advective problems, the accuracy of ELM solutions which are generated with accurate tracking algorithms deteriorate as the number of time steps increases. The assumption of exact tracking implies that there are no time truncation errors associated with the tracking step, and, since there is no diffusion in this problem, time truncation errors associated with equation (6) are irrelevant. Consequently, decreasing the time step does not reduce time truncation errors. ELM solutions deteriorate with decreasing time step, because interpolation errors are accumulated at at each time [Baptista 1982] and [Bentley, et. al. 1989] and, more time steps are required. Subsequently, as the results demonstrate, problems 2D through 2F are the most challenging for ELMs.

Appendix II. CD Forum Problem Set 4. The objective of the nine problems in this set is to transport a plume in a steady shear flow with constant diffusion. The velocity field is:

$$V_x = V_0 + \lambda y$$

$$V_y = 0 , \quad \text{(AII.1)}$$

where V_x and V_y are elements of the velocity vector, V_0 is the velocity along $y = 0$ and the shear, $\lambda = \dfrac{\partial V_x}{\partial y}$.

For the initial condition of a point source of mass M at $x = x_0$, $y = y_0$ and $t = 0$, the analytic solution for an infinite domain in which concentrations go to zero at infinity is [20]:

$$C(x,y,t) = \frac{M}{4\pi Dt(1+\lambda^2 t^2/12)^{\frac{1}{2}}}$$

$$\exp- \left[\frac{(x-\bar{x}-0.5\lambda yt)^2}{4Dt(1+\lambda^2 t^2/12)} + \frac{y^2}{4Dt} \right] , \quad \text{(AII.2)}$$

where $\bar{x} = x_0 + V_0 t$ and D is the diffusion coefficient.

The test problem initial conditions are the analytic solution of the problem evaluated at t_0. The initial mass for the test problems is:

$$M = 4\pi Dt_0(1+\lambda^2 t_0^2/12)^{\frac{1}{2}} , \quad \text{(AII.3)}$$

and, consequently, the initial peak concentration is one.

Eight of the problems are solved on a uniform grid, Grid 7:

$$x(i,j) = 200(i-1) \qquad i = 1,121 ,$$

$$y(i,y) = 200(j-1)-3400 \qquad j = 1,35 , \quad \text{(AII.4)}$$

where $x(i,y)$ and $y(i,j)$ are nodal coordinates and i,j are the nodal indices. An irregular grid is specified for one problem and it is called Grid 8:

$$x(i,j)-x(i-1,j) = 200-50\cos\frac{\pi(i-1)}{121} \qquad i = 2,121 ,$$

$$y(i,j+1)-y(i,j) = 200-50\cos\frac{\pi(j-1)}{35} \qquad j = 1,34 . \quad \text{(AII.5)}$$

The solutions of ARGOS 2D and ELMLQ were computed using the basic node spacing of 200. The node locations used by ELLESCO can be arrived at by replacing the values of 200 in equations (AII.4) and (AII.5) with 400, the 50 in equation (AII.5) with 100 and the index sweeps of $i = 1, 121$ with $i = 1, 61$ and $j = 1, 35$ with $j = 1, 18$. The initial concentration peaks are located at $x = 7200$, $y = 0$ on Grid 7 and $x = 7070.6$, $y = 0$ on Grid 8.

The details of the nine problems are found in Table 4. Problem Set 4 has an average grid Peclet number of $Pe = \dfrac{V_0 \Delta x}{D} = 10$ and average Courant number of $Co = \dfrac{V_0 \Delta t}{\Delta x} = 0.24$. The purpose of Problem Set 4 is to test the ability of methods to perform when diffusion is an important factor. Although ELMs are most advantageous for advective-dominated problems, they must also be able to perform adequately when significant amounts of diffusion are present. In contrast to Problem Set 2, time truncation errors are an issue in Problem Set 4, and problems 4G and 4F demonstrate the effects of increasing the time step increment. Other problems demonstrate the performance of the methods with greater and lesser shear, narrower and broader plumes, more diffusion and, finally, variable node spacings.

Appendix III. Error Measures. The L_2 error norm normalized by the mass is an integral measure of the overall error of the numerical solution.

$$
\phi_e = \frac{\left[\displaystyle\int_\Omega (\hat{C} - C^{ex})^2 \, d\Omega \right]^{1/2}}{\displaystyle\int_\Omega C^{ex} \, d\Omega} ,
\tag{AIII.1}
$$

where $\hat{C}$ is the numerical solution, C^{ex} is the analytic solution, and Ω is the domain. The value of ϕ_e for an exact solution is zero.

The error in the peak concentration normalized by the exact peak concentration is a point measure of amplitude errors in the numerical solution.

$$
\varepsilon_e = \frac{C^{ex}_{max} - \hat{C}_{max}}{C^{ex}_{max}} ,
\tag{AIII.2}
$$

where $\hat{C}_{max}$ and C^{ex}_{max} are the maximum values of the numerical and analytic solutions respectively. The measure ε_e is zero, greater than zero or less than zero for numerical solutions which are exact, overdamped or underdamped respectively.

The absolute value of the maximum negative concentration normalized by the maximum value of the analytic solution is a point measure of spurious oscillations caused by phase errors in the numerical solution.

$$
\Psi = abs \left[\frac{\hat{C}_{min}}{C^{ex}_{max}} \right] ,
\tag{AIII.3}
$$

where $\hat{C}_{min}$ is the most negative value of the numerical solution. The measure Ψ is zero for an exact numerical solution, and large values indicate phase errors in undamped higher spatial frequencies of the solution.

The total mass of the numerically computed concentration field normalized by the total mass of the analytic solution is an integral measure of mass conservation.

$$\mu_0 = \frac{\int\limits_{\Omega} \hat{C}\, d\Omega}{\int\limits_{\Omega} C^{ex}\, d\Omega}.$$ (AIII.4)

The measure μ_0 is one for an exact numerical solution.

References

1. G. F. Pinder and W. G. Gray, FINITE ELEMENT SIMULATION IN SURFACE AND SUBSURFACE HYDROLOGY, Academic Press, 1977.

2. F. M. Holly, Jr. and A. Preissmann, *Accurate calculation of transport in two dimensions*, J. HYDR. DIV., ASCE, HY11 (1977), pp. 1259-1277.

3. R. E. Ewing and T. F. Russel, *Multistep Galerkin methods along characteristics for convection-diffusion problems'*, in: ADVANCES IN COMPUTER METHODS FOR PARTIAL DIFFERENTIAL EQUATIONS, IV (1981), pp. 28-36.

4. S. P. Neuman, *A Eulerian-Lagrangian numerical scheme for the dispersion-convection equation using conjugate space-time grids*, J. COMP. PHYS., 41(1981), pp. 270-294.

5. J. Douglas, Jr. and T. F. Russel, *Numerical methods for convection-dominated diffusion problems based on combining the method of characteristics with finite element or finite difference procedures*, SIAM J. NUMER. ANAL., 19 (1982), pp. 871-885.

6. J. Glass and W. Rodi, *A higher order numerical scheme for scalar transport*, COMP. METHS. IN APPL. MECH. AND ENG., 31 (1982), pp. 337-358.

7. S. P. Neuman and S. Sorek, *Eulerian-Lagrangian methods for advection-dispersion*, FINITE ELEMENTS IN WATER RESOURCES (K.P. Mole, et.al., ed.), 4, 1982, pp. 14.41-14.68.

8. F. M. Holly, Jr. and T. Komatsu, *Derivative approximations in the two-point fourth order method for pollutant transport*, in "PROCEEDINGS OF THE CONF. ON FRONTIERS IN HYDR. ENG.", ASCE, MIT, Cambridge, 1983, pp. 349-355.

9. A. E. Baptista, E. E. Adams and K. D. Stolzenbach, EULERIAN-LAGRANGIAN ANALYSIS OF POLLUTANT TRANSPORT IN SHALLOW WATER, Ralph M. Parsons Laboratory, MIT, Rpt. No. 296., 1984.

10. F. M. Holly, Jr. and J. M. Usseglio-Polatera, *Dispersion simulation in two-dimensional tidal flow*, J. HYDR. DIV., ASCE, 110 (1984), pp. 905-926.

11. T. Komatsu, F. M. Holly, Jr., N. Nakashiki, and K. Ohgushi, *Numerical calculation of pollutant transport in one and two dimensions*, J. HYDROSCIENCE AND HYDR. ENG., 3 (1985), pp. 15-30.

12. A. E. Baptista, SOLUTION OF ADVECTION-DOMINATED TRANSPORT BY EULERIAN-LAGRANGIAN METHODS USING THE BACKWARDS METHOD OF CHARACTERISTICS, PhD Dissertation, M.I.T., 1987.

13. K. Toda and F. M. Holly, *Hybrid numerical method for linear advection- diffusion*, MICROSOFTWARE FOR ENG., 3 (1987), pp. 199-205.

14. L. R. Bentley, G. F. Pinder and I. Herrera, *Solution of the advective-dispersive transport equation using a least squares collocation, Eulerian-Lagrangean method*, NUMER. METH. PAR. DIF. EQ., 5 (1989), pp. 227-240.

15. L. R. Bentley, A. Aldama and G. F. Pinder, *Fourier analysis of the Eulerian-Lagrangian least squares collocation method'*, INT. J. NUMER. METH. FLUIDS, 11 (1990), pp. 427-444.

16. J. P. Laible and G. F. Pinder, *Least squares collocation solution of differential equations on irregularly shaped domains using orthogonal meshes*, NUMER. METH. PAR. DIF. EQ., 5 (1989), pp. 347-361.

17. A. Baptista, P. Gresho and E. Adams, *Instructions for the Convection Diffusion Forum of*

the VII Int'l Conf. on Comp. Meth. in Water Resources, Cambridge, Mass. (1988), personal communication.

18. J. A. Hoopes and D. R. F. Harleman, *Wastewater recharge and dispersion in porous media,* J. OF THE HYDR. DIV., ASCE, 93 (1967), pp. 51-71.

19. P. S. Huyakorn, P. F. Anderson, O. Guven and F. J. Molz, *A curvilinear finite element model for simulating two-well tracer tests and transport in stratified aquifers,* WAT. RES. RES., 22 (1986), pp. 663-678.

20. A. Okubo and M Karweit, *Diffusion from a continuous source in a uniform shear flow,* LIMN. AND OCEANO., 14 (1969), pp. 514-520.

CHAPTER 10

Finite Element Modelling of Surface Flow Problems*

Mehmet S. Cuhadaroglu**
David R. Maidment**
Linda J. Hayes†

Abstract. A two-dimensional implicit finite element model is used to solve the non-conservation form of the shallow water equations for overland flow on land surfaces. The model allows spatial variations in land characteristics and temporal changes in flow properties, namely water depth and velocity in the two horizontal directions. The paper also reviews alternative finite element methods to solve surface flow problems and the difficulties encountered in doing so.

Introduction. There are many types of hydraulic surface flow phenomena, for instance, open-channel flow, storm water runoff, dam-break flood flow and tidal flooding. An accurate prediction of surface flow is important and valuable in the field of engineering for the purposes of design, construction and operation of hydraulic structures and is also required in building site development for estimation of water quantity, quality, and flood damage.

Most computations of surface flow do not include any explicit characterization of the spatial and temporal distribution of flow velocity and depth, including those performed by standardized computer programs such as HEC-1 (1). Where the spatial distributions of velocity and depth are considered, they are usually considered only in one dimension, such as along the main channel of a river, and those computations are most often performed assuming steady flow, such as with the standardized computer program HEC-2 (2). Standardized programs are available from the US National Weather Service for unsteady, spatially varied flow in one dimension(3).

This paper presents a scheme for computation of depth and velocity in the two horizontal dimensions, using the finite element method. The governing equations of surface flow can be described by shallow water theory which is based on the nonlinear laws for conservation of mass and momentum. Due to the complexity of this problem and its geometry, it is impossible to solve the highly nonlinear hyperbolic system of equations analytically. Analytical

*This work was supported by the National Science Foundation, Grant No. MSM-8717452 and by The University of Texas at Austin.
**Department of Civil Engineering, University of Texas at Austin, Austin, TX 78712.
†Department of Aeorspace Engineering and Engineering Mechanics, University of Texas at Austin, Austin, TX 78712.

solutions are available only for simple geometries and if certain terms are neglected or linearized in the governing equations.

Various finite element methods have been developed for computations for flow in estuaries, rivers and flood plains (4,7) and the main differences among them lie in their time marching schemes, which can be grouped into two categories: namely, implicit schemes (5,8), which require procedures to solve simultaneous algebraic equation systems, and explicit schemes (9,12), which do not require these procedures. The present work differs from those mentioned above in being primarily directed at computing the dynamics of overland flow on land surfaces, as distinct from flow in streams and estuaries. However the numerical method presented here is similar to implicit schemes that were previously developed by other authors (5,8).

Governing equations. The Continuity and Navier-Stokes equations for an incompressible fluid in an Eulerian coordinate system are:

$$\nabla . v = 0 \qquad (1)$$

$$\frac{\partial v}{\partial t} + (v . \nabla) v = - \frac{\nabla P}{\rho} + v \nabla^2 v + f \qquad (2)$$

where

$$v = \begin{pmatrix} u \\ v \\ w \end{pmatrix}$$

are the x, y, z components of the velocity vector and P, ρ, and v are pressure, density and kinematic viscosity respectively. Figure 1 shows the coordinate system employed. And

$$f = \begin{pmatrix} f_x \\ f_y \\ f_z \end{pmatrix}$$

are external (gravitational) body forces in the x, y and z directions. Figure 1 shows the coordinate system employed.

The governing equations of surface flow can be described by shallow water theory, in which the three dimensional flow problem, represented by continuity eq.(1) and Navier - Stokes eq.(2) equations, is approximated by averaging the flow properties in the vertical direction and assuming the vertical pressure distribution is hydrostatic. Two types of shallow water equations can be used in these analyses, the non-conservation form, which employs velocity and water depth as the dependent variables, or the conservation form, which uses discharge and water surface elevation as the dependent variables.

After depth-averaging, the vector form of the shallow water equations in non-conservation form is:

$$\frac{\partial h}{\partial t} + (v . \nabla) h + h (\nabla . v) = i_0 \qquad (3a)$$

$$\frac{\partial u}{\partial t} + (v . \nabla) u = - g \frac{\partial \eta}{\partial x} - g \frac{\sqrt{u^2 + v^2}}{hc^2} u + (\nabla . \varepsilon \nabla) u \qquad (3b)$$

$$\frac{\partial v}{\partial t} + (v . \nabla)v = -g\frac{\partial \eta}{\partial y} - g\frac{\sqrt{u^2 + v^2}}{hc^2}v + (\nabla . \varepsilon\nabla)v \tag{3c}$$

where u and v are now the vertically averaged flow velocities, and

ε: Eddy viscosity (L^2T^{-1}): $\varepsilon = C_v h\sqrt{u^2 + v^2}$

C_v: Adjustable eddy viscosity parameter

h: Local depth (L)

i_0:   Excess rainfall (LT^{-1})

u,v: Velocity components in x and y directions (LT^{-1})

η: Water surface coordinate (L): $\eta = \zeta + h$

ζ: Ground surface coordinate (L)

g: Acceleration of Gravity (LT^{-2})

t: Time (T)

c: Chezy coefficient $(\sqrt{L}T^{-1})$

The boundary conditions that were used to derive the governing equations are no slip and no seepage on the channel bed and atmospheric pressure and no shear stress on the free surface. These boundary conditions are given below:

On the bed:

$$z = \zeta(x, y, t)$$
$$u(x, y, \zeta, t) = 0$$
$$v(x, y, \zeta, t) = 0$$
$$w(x, y, \zeta, t) = 0$$

$$\frac{Dz}{Dt} = \frac{D\zeta}{Dt} = w(x, y, \zeta, t)$$

$$w(x, y, \zeta, t) = \frac{\partial \zeta}{\partial t} + u(x, y, \zeta, t)\frac{\partial \zeta}{\partial x} + v(x, y, \zeta, t)\frac{\partial \zeta}{\partial y}$$

And under the boundary conditions on the bed given above:

$$\frac{\partial \zeta}{\partial t} = 0 \tag{4}$$

As the free surface:

$$z = \eta(x, y, t)$$
$$P = P_0 \text{ (or simply } P = 0)$$

$$\frac{Dz}{Dt} = \frac{D\eta}{Dt} = w(x, y, \eta, t)$$

$$w(x, y, \eta, t) = \frac{\partial \eta}{\partial t} + u(x, y, \eta, t) \frac{\partial \eta}{\partial x} + v(x, y, \eta, t) \frac{\partial \eta}{\partial y} - i_0 \tag{5}$$

The definition of depth-averaging assumes the flow is predominantly horizontal, that is the flow velocity in both x and y directions is comparable and is much larger than the vertical velocity component so that the flow properties can be averaged in the vertical direction, and the vertical pressure distribution is hydrostatic. These conditions are stated follows:

$$U = \frac{1}{h} \int_{\zeta}^{\eta} u \; dz \tag{6a}$$

$$V = \frac{1}{h} \int_{\zeta}^{\eta} v \; dz \tag{6b}$$

$$P = P_0 + \rho g(\eta - \zeta) \tag{7}$$

Finite element modelling. An advantage of the finite element method is that it can be easily applied to solve problems on complicated domains. The computations conducted for implicit schemes are usually more stable than those for explicit schemes (5,8), although the computational times for explicit schemes are usually less. Viscous forces are often neglected in the governing equations to save computation time, but realistically the flow is dominated by viscous effects as well as bottom and side wall friction (11,12). The inviscid equations are irrotational and do not provide the significant eddies and flow separations that occur in most shallow water flow problems. This paper uses the complete form of shallow water equations without neglecting the viscous forces.

Inviscid flow equations can, however, play a productive role in the numerical modelling of real flows for a different reason. The solution of inviscid equations does not provide flow separation or other evidence of rotational behavior, regardless of the geometry that is provided. One can also argue that for most shallow water problems, walls can be adequately defined as slip boundaries, which neglects their viscous forces because bottom friction is the dominant source of friction. The bottom friction effect is added to the equations whether there is viscous effect in the fluid or not. Practically speaking, when the flows encounter the shallow water near boundaries, velocities are significantly reduced due to bottom friction effects. Also, since the longitudinal and transverse dimensions are very large, compared to the depth, the effect of viscous forces around boundaries due to no-slip boundary conditions is negligible compared to inertial forces and pressure over the whole solution domain.

To derive the finite element model, the domain of interest is divided into many small triangles, or rectangles. Then using variational principles, the non-conservation form of eqs.(3), are recast in an equivalent integral form by exchanging the differentiation between a test function, W, and the dependent variables, H, U, V in our case(13). An implicit finite element formulation is employed, using a standard Galerkin approach in space and a backward differencing in time which employs velocity and water elevation as dependent variables. The following coupled system of equations are solved to give H^{i+1}, the water depth; U^{i+1}, the velocity in x - direction; and V^{i+1}, the velocity in y - direction at time i+1.

Continuity equation:

$$\int_{\Omega} \left(\frac{H^{i+1}}{\Delta t}W + U^i \frac{\partial H^{i+1}}{\partial x}W + V^i \frac{\partial H^{i+1}}{\partial y}W + H^{i+1}\frac{\partial U^i}{\partial x}W + H^{i+1}\frac{\partial V^i}{\partial y}W \right) d\Omega$$

$$= \int_{\Omega} \frac{H^i}{\Delta t}W\, d\Omega + \int_{\Omega} I_0^{i+1}W\, d\Omega \tag{8}$$

Momentum equation in x - direction:

$$\int_{\Omega}\left(\frac{U^{i+1}}{\Delta t}W + \frac{g\sqrt{U^{i2}+V^{i2}}}{H^i\,C^2}U^{i+1}W + \varepsilon^i\left(\frac{\partial U^{i+1}}{\partial x}\frac{\partial W}{\partial x} + \frac{\partial U^{i+1}}{\partial y}\frac{\partial W}{\partial y} \right) + U^i\frac{\partial U^{i+1}}{\partial x}W \right.$$

$$\left. + V^i\frac{\partial U^{i+1}}{\partial y}W \right) d\Omega = \int_{\Omega} \frac{U^i}{\Delta t}W\, d\Omega - \int_{\Omega} g\frac{\partial \eta}{\partial x}W\, d\Omega + \int_{\partial \Omega} \varepsilon^i\frac{\partial U^{i+1}}{\partial n}W\, ds \tag{9}$$

Momentum equation in y - direction:

$$\int_{\Omega}\left(\frac{V^{i+1}}{\Delta t}W + \frac{g\sqrt{U^{i2}+V^{i2}}}{H^i\,C^2}V^{i+1}W + \varepsilon^i\left(\frac{\partial V^{i+1}}{\partial x}\frac{\partial W}{\partial x} + \frac{\partial V^{i+1}}{\partial y}\frac{\partial W}{\partial y} \right) + U^i\frac{\partial V^{i+1}}{\partial x}W \right.$$

$$\left. + V^i\frac{\partial V^{i+1}}{\partial y}W \right) d\Omega = \int_{\Omega} \frac{V^i}{\Delta t}W\, d\Omega - \int_{\Omega} g\frac{\partial \eta}{\partial y}W\, d\Omega + \int_{\partial \Omega} \varepsilon^i\frac{\partial V^{i+1}}{\partial n}W\, ds \tag{10}$$

where

$$t \in (0,T)$$

$$\Delta t = T/n$$

$$\Delta t > 0$$

$$t^i = i\Delta t$$

$$H^i = H(.,\, t^i)$$
$$U^i = U(.,\, t^i)$$

$$V^i = V(.,\, t^i)$$

$T : total\ time$

$\Delta t : time\ step\ length$

$n : total\ number\ of\ calculation\ steps$

$i = 0\ corresponds\ to\ the\ initial\ data$

and W is a test function. This global system of equations with proper boundary and initial conditions imposed is solved directly for the dependent variables.

Numerical examples. The implicit finite element code using depth-averaging methods was first used to model overland flow on a land surface near Austin, Texas for which the mesh is given in Figure 2. The spatial and temporal variations of depth and velocity of a possible overland flow induced by a storm are important for future building site development. The domain was manually divided into triangles with a spacing range of 50 - 500 ft depending upon the local slope. Constant inflow boundary conditions and stationary water surface elevation were used as initial conditions. The value of Chezy coefficient, and Manning's coefficient n, accounting for the bottom roughness were 15 and 0.1 respectively (a grassy natural surface was assumed). For the given inflow conditions, it appears that the model could successfully analyze the overland flow in terms of directions and magnitudes of the velocities, shown in Figure 3, and water depths, shown in Figure 4, depending upon the topography of the land.

Natural channels are characterized by highly irregular cross-sections accompanied by abrupt contractions and/or expansions in width and/or depth. To see the effects of such geometric complexities, several special cases were tested. A sudden contraction of width in a channel such as the one shown in Figure 5 should cause a sudden change in water properties around the contraction. It becomes necessary for the model to capture those changes for an accurate flow analysis. Also a sudden elevation change in channel bottom which can be confronted when analyzing flow over a rough surface becomes another important case of interest. Figures 8a and 8b show a possibble such a case. For the possibility of using this model for storm water analysis, flow around sharp objects, such as a barrier shown in Figure 11, became another important case to analyze.

For the case where a sudden contraction of width occurs in Figure 5, uniform inflow at the entrance and no normal flow at walls were used as boundary and initial conditions, respectively. The flow field and water elevation results are given in Figures 6 and 7. The sudden changes in water properties, such as the high water depths in the lower corner, near point D, and acceleration of the flow at that point, were detected quite well.

The results for a straight channel with a bump in the middle and inclined bed elevation in the direction of the channel are given in Figures 8a, 8b, 9 and 10 which show the flow field and water elevations respectively. A uniform flow at the entrance was assumed. The flow slows down and water builds up while approaching the bump. As it flows over the bump, the elevation drops and the velocity increases.

The flow field results and water elevations for a straight channel with a barrier in the middle and inclined bed elevation in the lateral direction are given in Figure 12 and 13. A deceleration of flow before the barrier to a stagnation point at the leading edge and water build up along with an acceleration as it flows around the barrier are the main observations.

The main difficulty encountered in studying these test cases was the non-linear instabilities in the solution due to sudden changes in flow properties. In order to overcome these instabilities, some order of artificial viscosity, about as much as the viscosity of the momentum equation, was added to the depth averaged continuity equation. Also it may be possible by using very high order artificial viscosity and very fine mesh to analyze

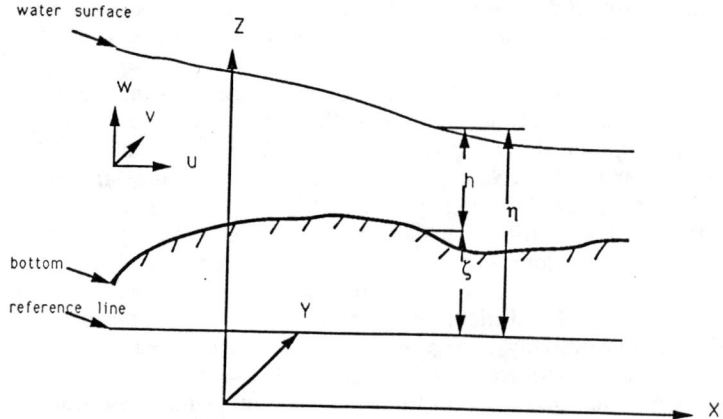

Figure 1 - The coordinate system

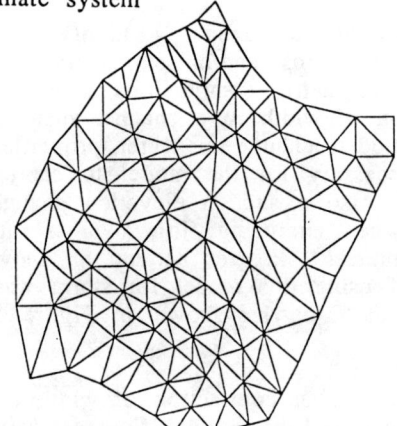

Figure 2 - The mesh of overland flow on a land surface near Austin,

Texas

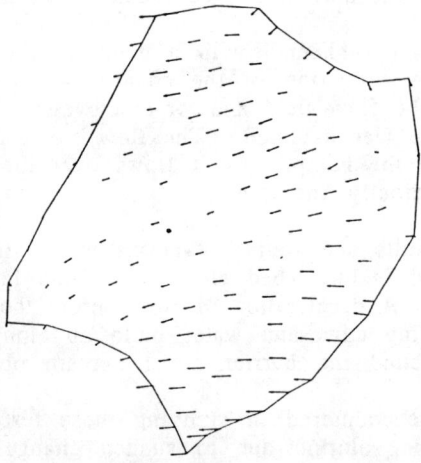

Figure 3 - The flow field of overland flow on a land surface near Austin,

Texas

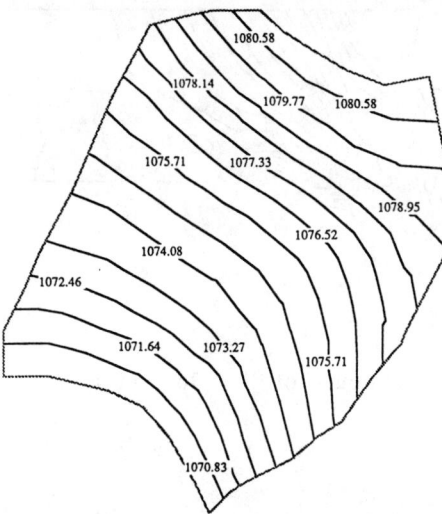

Figure 4 - The water contours of overland flow on a land surface near

Austin, Texas

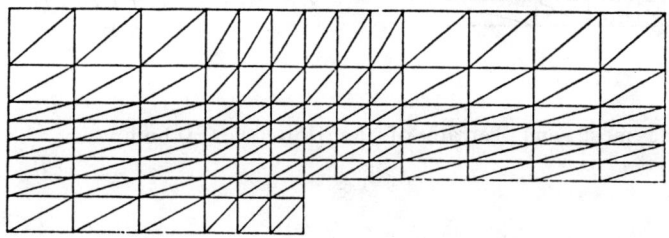

Figure 5 - The mesh of the flow in a contracting channel

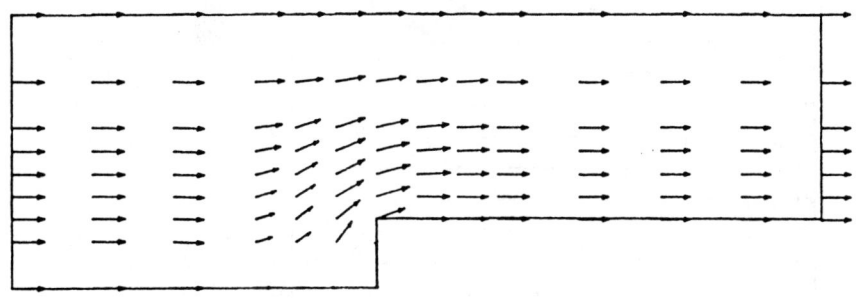

Figure 6 - The flow field of the flow in a contracting channel

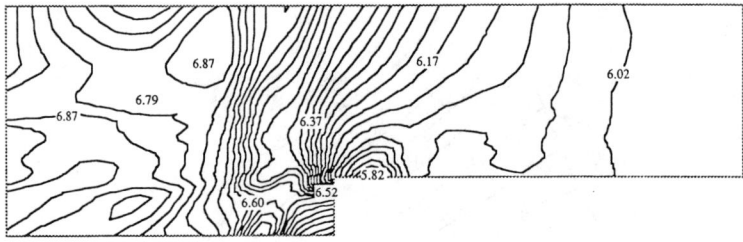

Figure 7 - The water contours of the flow in a contracting channel

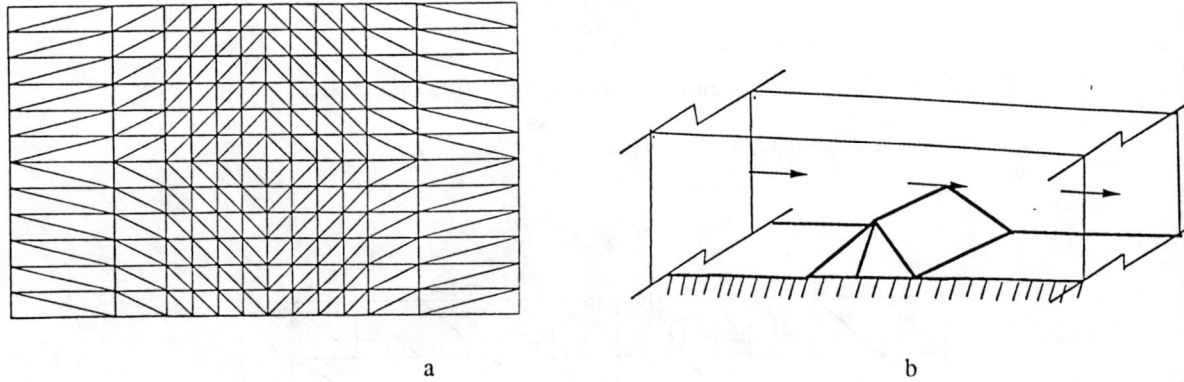

a   b

Figure 8 - The mesh and Terrain of the flow over a bump

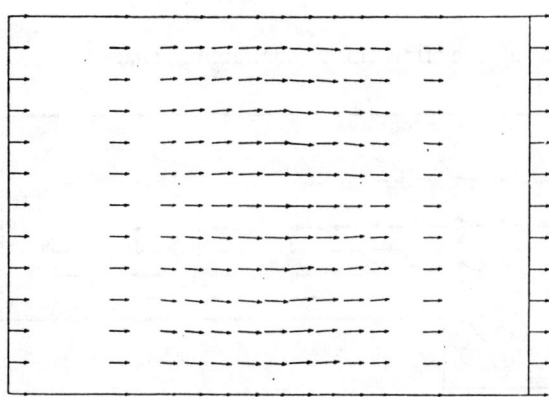

Figure 9 - The flow field of the flow over a bump

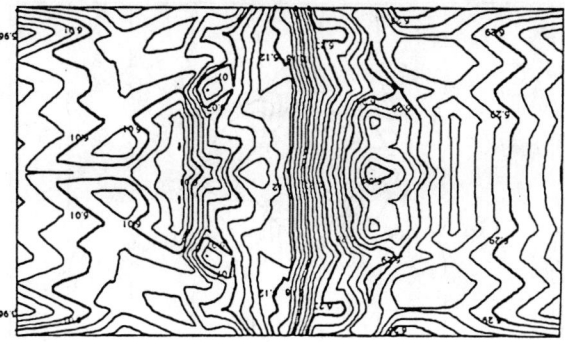

Figure 10 - The water contours of the flow over a bump

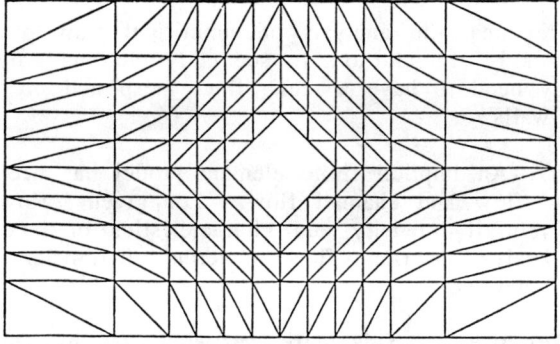

Figure 11 - The mesh of the flow around a barrier

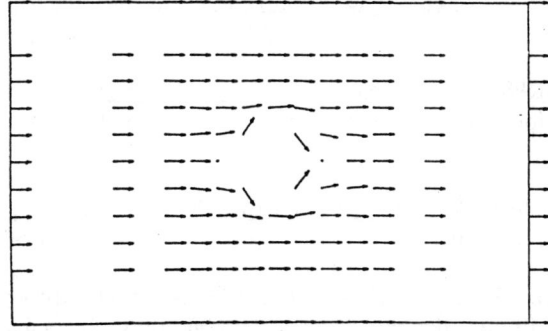

Figure 12 - The flow field of the flow around a barrier

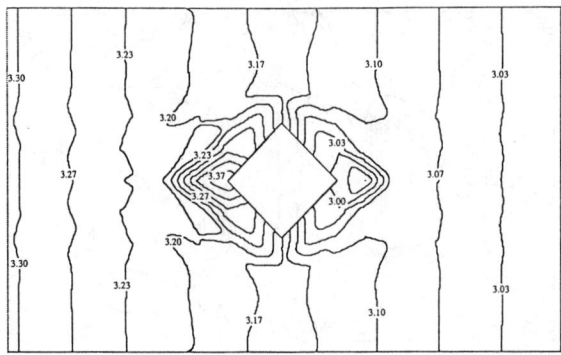

Figure 13 - The water contours of the flow around a barrier

supercritical flow regimes, the analysis of subcritical flow problems is more
realistic by this method.   In addition, to see eddy motions around sharp
geometries, one may need to have a much finer mesh and no - slip boundary
conditions on the walls.

Conclusion.   An implicit finite element model has been used for the
analysis of overland flow and channel flow.   For given inflow and/or outflow
conditions, excess rainfall intensity and characteristics of land, the model
here was used to analyze overland flow in terms of transient-temporal flow
patterns and water depths.

Acknowledgements.   This work was supported by The National
Science Foundation, Grant No. MSM-8717452 and by The University of Texas at
Austin.   The authors also would like to thank Dr. S. Wright and Mr. N. Jones for
their help during the course of this work.

REFERENCES

1.   U.S. ARMY CORPS OF ENGINEERS, HEC-1, Flood hydrography package, User's
Manual, Hydrologic Engineering Center, 609 Second St. Davis CA, Revised
Edition, 1987.

2.   U.S. ARMY CORPS OF ENGINEERS, HEC-2, Water surface profiles, User's
Manual, Hydrologic Engineering Center, 609 Second St. Davis CA, Revised
Edition, 1976.

3.   D. L. FREAD, Channel routing, in Hydrological Forecasting ed. by M. G
Anderson and T. P. Burt, Wiley, New York, 1985, pp. 437-503.

4.   J. F. COCHET, G. DHATT, G. HUBERT and G. TOUZOT, River and estuary flows by
a new penalty finite element, Finite element flow analysis, (1982), pp. 563-
570.

5.   T. Y. SU and S. Y. WANG, Depth-averaging models of river-flow, Finite
elements in water resources, (1980), pp. 5.223-5.235.

6.   J. K. LEE, Two-dimensional finite element analysis of hydraulic effect of highway bridge fills in a complex flood plain, Finite elements in water resources, (1980), pp.6.3-6.23.

7.   A. J. BAKER and M. O. SOLIMAN, On the accuracy and efficiency of a finite element algorithm for hydrodynamic flows, Finite elements in water resources, (1982), pp.2.39-2.56.

8.   A. J. BAKER and M. O. SOLIMAN, Analysis of a finite element algorithm for numerical predictions in water resources research, Finite elements in water resources, (1980), pp.1.40-1.55.

9.   M. KAWAHARA, N. TAKEUCHI and T. YOSHIDA, Two-step explicit finite element method for tsunami wave propagation analysis, Int. J. Num. Meth. Engng, 12, (1978), pp. 331-351.

10. M. KAWAHARA, S. NAKAZAWA, S. OHMORI and T. TAKAGI, Two-step explicit finite element method for storm surge propagation analysis, Int. J. Num. Meth. Engng, 15, (1980), pp.1129-1148.

11. J. PERAIRE, O. C. ZIENKIEWICZ and K. MORGAN, Shallow water problems: A general explicit formulation, Int. J. Num. Meth. Engng, 22, 1986, pp.547-574.

12. R. LOHNER, K. MORGAN and O. C. ZIENKIEWICZ, The solution of non-linear hyperbolic equation systems by the finite element method, Int. J. Num. Meth. Engng, 4, (1984), pp.1043-1063.

13. E. B. BECKER, G. F. CAREY and J. T. ODEN, Finite elements an introduction volume I, Prentice-Hall International, Inc., 1981.

CHAPTER 11

Forward and Inverse Modeling of the Advection-Diffusion Equation in the Presence of Sharp Fronts*

Manfred Koch**
Gangpeng Zhang**

Abstract The advection-diffusion equation (ADE) governs a variety of transport processes in various scientific disciplines. The parabolic/hyperbolic nature of the ADE frequently gives rise to the development of sharp fronts which are difficult to resolve by standard FD or FE methods. Techniques such as 'upwinding' reduce the numerical oscillations but introduce artificial dissipation so that they are inadequate for the modeling of sharp breakthrough-curves (BTC). These often arise in the presence of nonlinear adsorption isotherms or for multiphase flow in a porous media. Several classical FD and FE methods have been applied to a strongly advection dominated ADE and were found to be unsatisfactory. Thus we developed a new method of lines (MOL) based on an adaptive ODE solver for the forward modeling of the 1D ADE. Two variants of time-integration were investigated. One is based on a first order backward Euler integation (first-order MOL) and the other on a second order Taylor truncated time-discretization (Taylor MOL). In addition to the classical linear parabolic/hyperbolic ADE, the MOL has been applied to a nonlinear chromatographic plug-flow model and to the nonlinear two-phase Buckley-Leverett flow. The results show that both versions of the new adaptive MOL method are unconditionally stable and essentially dispersion-free. The first-order MOL shows some diffusion that can be minimized by taking sufficiently small time steps. The Taylor MOL, on the other hand, exhibits much less diffusion and allows the use of larger time steps. However, the more delicate balancing of spatial- and time derivatives makes the Taylor MOL more prone to numerical dispersion. In the case where an analytical solution of the ADE can be obtained, we have performed inverse modeling of various transport parameters from a chromatographic elution profile using two nonlinear optimization techniques, namely the Levenberg-Marquardt method and the BFGS-variable metric method. The latter technique proved to have a greater radius of convergence than

*The present work has been sponsored by the Department of Energy through a Contract No. DE-FC05-85ER250000. Additional support has been provided by the Florida State University through the allocation of supercomputer resources on the ETA10.
**Supercomputer Computations Research Institute, Geophysical Fluid Dynamics Institute, and Department of Geology, Florida State University, Tallahasse, FL 32306.

the first one. The theoretical model obtained with the BFGS method was able to fit the observed chromatogram very satisfactorily.

1. INTRODUCTION AND MATHEMATICAL FORMULATION

Advection-diffusion governed transport problems play an important role in many scientific areas such as fluid dynamics (Fletcher, 1988) convective heat transfer (Gebhart, et al., 1988), solute transport processes (Bear, 1972; 1979), chemical chromatography (Conder and Young 1979) and petroleum reservoir modeling (Allen, et al., 1988a; Wheeler, 1988).

All of these physical transport processes are governed by the advection-diffusion equation (ADE) which is a second order nonlinear parabolic/hyperbolic partial differential equation (PDE) that expresses the mass conservation of the physical quantity $u = u(x,t)$ to be transported.

The ADE can be written mathematically for the 1D case in the following form (Lapidus and Pinder, 1982):

$$\frac{\partial u}{\partial t} = f(x, u, u_x, u_{xx}) \tag{1.1a}$$

or in explicit form as

$$\partial u / \partial t + F(u, x)_x = (D(u, x)u_x)_x + q(x) \tag{1.1b}$$

or

$$\frac{\partial u}{\partial t} + f_{fl}(u, x)u_x = (D(u, x)u_x)_x + q(x) \tag{1.1c}$$

where x is defined in the interval (a, b), $f_{fl}(u, x) = \partial F(u, x)/\partial u$ is a nonlinear flow-function describing the advection, and $D(u, x)$ is the diffusion function. The source term is denoted as q. Henceforward we assume q to be zero. For the special case that $F(u, x)$ is linear, i.e., $h(u, x) = v = $ constant and $D(x, u) = \epsilon = $ constant, one obtains

$$\frac{\partial u}{\partial t} + v\frac{\partial u}{\partial x} = \epsilon u_{xx} \tag{1.1d}$$

The parabolicity/hyperbolicity of the PDE (1.1b) to (1.1d) is determined by the ratio of the diffusion function D or the diffusion coefficient ϵ, respectively, to the flow function f_{fl} or the advection velocity v, respectively. Depending on the size of the ratio ϵ/v in (1.1d) the transport problem will be either diffusion or convection/advection dominated.

The first case, $\epsilon/v \gg 1$ and, in particular, the limiting case, $v = 0$, represent a dissipative transport problem and is governed by a parabolic PDE which very often can be solved by analytical means and is readily solvable numerically by either a finite-difference or finite-element method (see e.g. Lapidus and Pinder, 1982; Allen, et al., 1988b).

In the second case, $\epsilon/v \ll 1$, one obtains an advection dominated transport problem which for the limiting case $\epsilon = 0$ converts to a fully hyperbolic PDE and thus is written as

$$\frac{\partial u}{\partial t} + v\frac{\partial v}{\partial x} = 0 \tag{1.2a}$$

which is the special case of a hyberbolic conservation law (Sod, 1985)

$$\frac{\partial u}{\partial t} + \frac{\partial F}{\partial x} = 0 \qquad (1.2b)$$

or

$$\frac{\partial u}{\partial t} + f_{fl}(u, x) u_x = 0 \qquad (1.2c)$$

which are the special forms of (1.1b) and (1.1c) with $D = 0$.

In the advection dominated case, $\epsilon/v \ll 1$, the solution of the ADE (1) whether obtained analytically—which in most cases is only possible for a certain range of transport parameters and boundary conditions (Bear, 1979; Van Genuchten and Alves, 1982)—or numerically, is particularly intricate. In fact, advection dominated transport flow problems often give rise to sharp fronts in the solution u, i.e., steep **breakthrough curves (BTC)** are generated. For the pure hyberbolic case (1.2) of zero diffusion, a shock in the solution profile can even be produced under certain physical conditions—which are mathematically epitomized in the nature of the function F or f_{fl})— (Allen, et al., 1988a).

In all of these cases it is very difficult to resolve the sharp front correctly. In fact, the numerical solution will often exhibit **numerical dispersion** and/or **numerical diffusion** (Gray and Pinder, 1976; Allen at al., 1988b; Fletcher, 1988) which makes the solution practically worthless. An illustration of these effects is shown in Fig. 3.1 which will be discussed later.

Modeling of a physical transport process governed by the ADE can be done in two ways:

a) **Forward modeling:**

The relevant transport parameters in the ADE in conjunction with a set of initial and boundary conditions are specified and the PDE is solved by an analytical or numerical technique. To comply with observed BTC's the free parameters in the ADE are adjusted in a 'trial-and-error' procedure until the time history of the BTC is matched. The success of the forward modeling depends on the ability to solve the ADE properly given the numerical problems in the presence of sharp fronts.

In the present paper we will focus on the implementation of a new numerical adaptive method of lines to solve the forward problem for the advection dominated ADE (1.1b) for linear and nonlinear flow functions, $F(u, x)$, and for small physical diffusion, $D(u, x)$. For the extreme case of zero diffusion we will solve the pure hyberbolic equation (1.2). The need for a better numerical technique arose after we encountered all of the above mentioned numerical problems of dispersion and diffusion in existing library codes such as IMSL and NAG and in various finite difference and finite element methods as discussed in Section 3.1.

b) **Inverse modeling:**

The transport parameters are determined directly from an observed BTC, i.e., knowing the solution of the ADE, the objective is to compute the input parameters of the model. As such, inverse modeling can be viewed as solving a nonlinear multiparameter estimation problem through a time history matching of the observed output of the model. While parameter estimation of problems governed by regular parabolic PDE's, such as for heat conduction (Beck and Arnold, 1977) for diffusion controlled solute transport or for transient groundwater flow (Carrera and Neuman,

1986; Yeh, 1986) have been applied with success in recent years, inverse modeling of the advection dominated ADE has not yet found its due place in the literature. To a large extent this is a consequence of the mathematical and numerical difficulties encountered when sharp fronts are present in the solution of the ADE. Only in cases where analytical solutions of the ADE can be found, such as for the special case (1.1d) of the ADE, is inverse modeling, in principle, feasible. Nevertheless, as has been shown recently by Wagner and Gorelick (1986) and Knoppman and Voss (1987) through sensitivity analysis, inherent problems such as **identifiability, uniqueness** and **stability** of the inverted model do still present large obstacles for routine inverse modeling. We will discuss some of these problems in Chap. 4 when inverting an analytical BTC by a nonlinear optimization techniques.

To tackle the inverse problem we have developed two nonlinear iterative least-squares techniques based on the **Levenberg-Marquardt (LM) method** (Levenberg 1944; Marquardt, 1963) and a **BFGS-variable metric technique** (Fletcher, 1980). Examples for the inversion of a chromatographic elution BTC, which under the assumption of linear adsorption can been modeled by an analytical solution of the ADE, show the versatility and limitations of the methods.

These inversion methods are presently being implemented in the new numerical method of lines to allow the inversion for nonlinear flow functions as they apply in general adsorption models or multiphase flow as will be discussed in Section 2.

2.   PHYSICAL AND MATHEMATICAL INTERPRETATION OF THE ADE FOR SPECIAL CASES

Before discussing some special forms of the ADE as they are modeled in the present paper, two additional physical applications of the ADE (1.1b) are to be noted, namely **convective heat transfer** (Gebhardt, et al., 1988) and **Burgers' equation** (Fletcher, 1988) used in high Reynold's number Fluid Dynamics. In the latter case the flow function $F(u, x)$ in Eq. (1.1b) is nonlinear and equal to $F(u) = 1/2u^2$, while D denotes the viscosity. The special case of the inviscid hyperbolic Burgers' equation (1.2b) is obtained for $D = 0$. The numerical solution of this equation is known to be difficult due to the development of shocks (Sod, 1985). The numerical solution of Burgers' equation using the new method of lines presented in Section 3 is presented in (Koch, 1992a).

2.1.   Single phase flow of miscible solutes through a porous media

This area includes problems of:

- Contaminant transport in groundwater flow (Bear, 1972; 1979).

- Transport of chemicals through adsorbing beds and chromatographic systems (Lapidus and Amundson, 1952; Conder and Young, 1977).

For both cases, for which we will discuss the development of sharp fronts in more detail, u in (1.1) denotes the concentration C of the solute.

The general ADE including adsorption for the concentration $C_m(x, t)$ of the mobile aqueous phase and the concentration $C_s(x, t)$ of the adsorbed (stationary) phase of a transported miscible solute can be derived from a mass-balance consideration (Bear, 1979). Assuming that D and v are constant and omitting chemical decay one obtains for the 1D-case the following ADE:

$$\frac{\partial C_m}{\partial t} + k_s \frac{\partial C_s}{\partial t} + \frac{v}{\phi} \frac{\partial C_m}{\partial x} = D \frac{\partial C_m^2}{\partial x^2} \tag{2.1}$$

where $k_s = (1 - \phi)/\phi \rho_b$, where ϕ is the porosity of the media and ρ_b its bulk density. The other quantities are as above.

The relationship between C_s of the adsorbate and the concentration C_m of the aqueous phase depends on the kinetics of the adsorption process. In the case of equilibrium (instantaneous) adsorption the isotherm has the general form

$$C_s = F_{ad}(C_m). \tag{2.2}$$

With the equilibrium isotherm (2.2) one obtains for (2.1):

$$\left(1 + k_s \frac{\partial F_{ad}(C_m)}{\partial C_m}\right) \frac{\partial C_m}{\partial t} + \frac{v}{\phi} \frac{\partial C_m}{\partial x} = D \frac{\partial^2 C_m}{\partial x^2} \tag{2.3}$$

or

$$\frac{\partial C_m}{\partial t} + f_{ad}(C_m) \frac{\partial C_m}{\partial x} = D_s(C_m) \frac{\partial^2 C_m}{\partial x^2} \tag{2.4}$$

with

$$f_{ad}(C_m) = \frac{v}{\phi \left(1 + k_s \frac{\partial F_{ad}(C_m)}{\partial C_m}\right)} \tag{2.5}$$

and

$$D_s(C_m) = \frac{D}{\left(1 + k_s \frac{\partial F_{ad}(C_m)}{\partial C_m}\right)} \tag{2.6}$$

Note that (2.4) again has the general form of (1.1b) where the advective flow function $f_{ad}(C_m)$ (2.5) is now proportional to the inverse of the slope of the adsorption isotherm.

In chemical chromatography it is often assumed that the diffusion coefficient D is negligible against the advective flow function f (Conder and Young, 1977). In such a case one gets the so called **chromatographic plug-flow model**:

$$\frac{\partial C_m}{\partial t} + f_{ad}(C_m) \frac{\partial C_m}{\partial x} = 0 \tag{2.7}$$

which is the special form of the nonlinear hyperbolic equation (2.2) for a pure advective adsorptive transport process.

The behavior of the solution of (2.4) and (2.7) depends strongly on the nature of the isothermal relationship (2.2) between the mobile and stationary phases C_m and C_s, respectively. The most commonly used isotherms is
  Freundlich's isotherm

$$C_s = K_d C_m^{1/n} \tag{2.8}$$

where K_d and n are the distribution coefficient and an experimentally determined exponent, respectively (Bear, 1979).

The nonlinear functional relationship (2.8) reduces for the special case $n = 1$ to a linear isotherm, i.e., $C_s = K_d C_m$ and, using Eqs. (2.5) and (2.6), one gets for the ADE the following expression:

$$\frac{\partial C_m}{\partial t} + \frac{v}{R}\frac{\partial C_m}{\partial x} = \frac{D}{R}\frac{\partial C_m^2}{\partial t^2} \tag{2.9}$$

where the retardation factor R is given by $R = 1 + (1 - \phi)/\phi\rho_b K_d$. Eq. (2.9) is mathematically identical to Eq. (1.1b) with v being replaced by v/R and D by D/R which shows that linear adsorption just slows down the BTC by the factor R.

While linear adsorption ($n = 1$) does not change the shape of the BTC or a chromatographic elution profile it is known that for a general nonlinear isotherm ($n \neq 1$) (as described by Eq. 2.8) the shape of an initially symmetric elution peak in a chromatographic column will be skewed in a certain direction. For an exponent $n < 1$ in Eq. (2.8) the isotherm is concave upward which results in a steepening of the leading front while the tail of the pulse is trailing. For $n > 1$ the isotherm is concave downward and the situation is reversed such that the peak has now a sharpened tail while the front is flattened (Conder and Young, 1977; Allen, et al., 1988a). For the case of the hyperbolic plug-flow model (2.7) in both of these situations a shock might eventually occur. The presence of physical dispersion might, however, mitigate this situation to happen. A numerical simulation of the pure plug-flow model (2.7) using the new method of line (Section 3.3.3) is presented in the Fig. 3.5 and will be discussed in details then.

2.2. Multiphase flow through a porous media

The general ADE (1.1b) can be used for the description of **multiphase flow** occurring in:

- **Reservoir modeling** (Black oil models) (Azia and Settari, 1979; Allen, et al., 1988a; Parker, 1989)
- **Two-phase flow** through the partially unsaturated zone (Bear, 1972; Parker, 1989)

For immiscible multiphase flow composed of either water/oil in reservoir modeling or of water/air for the partially saturated soil zone one can derive a 1D ADE for the saturation S for one of the two phases (the wetting phase $S = S_w$), from the continuity equation and Darcy's flow equation (Bear, 1979)

$$\phi\frac{\partial S_w}{\partial t} + f_{BL}(S_w)\frac{\partial S_w}{\partial x} = -\frac{\partial}{\partial x}\left(D(S_w)\frac{\partial S_w}{\partial x}\right) \tag{2.10}$$

with

$$f_{BL}(S_w) = v\frac{\partial F_{BL}}{\partial S_w} \tag{2.11}$$

where

$$F_{BL}(S_w) = \frac{k_w(S_w)}{k_w(S_w) + k_{nw}(S_{nw})} \tag{2.12}$$

and

$$D(S_w) = F_{BL}(S_w)k_{nw}\frac{\partial p_c}{\partial S_w}. \tag{2.13}$$

In Eqs. (2.10) to (2.13) ϕ is the porosity of the media, v is the flow velocity, k_w and k_{nw} are the nonlinear wetting and nonwetting permeabilities, respectively, and $D(S_w)$ is the nonlinear diffusivity coefficient due to the capillary pressure p_c.

Eq. (2.10) is identical to (1.1c) with the flow function $F(u)$ being replaced by the general Buckley-Leverett function F_{BL}. Eq. (2.10), which is the most general form of the **Buckley-Leverett flow equation**, is also formally identical to the nonlinear adsorption ADE (2.4).

The parabolicity/hyberbolicity of Eq. (2.10) is given by the ratio of the capillary diffusivity D to the Buckley-Leverett flow function F_{BL}. What is usually called the Buckley-Leverett equation refers to the case that the effect of the capillary pressure is neglected (Buckley and Leverett, 1942). Thus one obtains the classical **Buckley-Leverett (BL) equation**

$$\phi\frac{\partial S_w}{\partial t} + f_{BL}(S_w)\frac{\partial S_w}{\partial x} = 0 \qquad (2.14)$$

which is hyberbolic in nature.

In a manner similar to the adsorption plug-flow model (2.7), the behavior of the solution of the BL equation (2.14) will depend on the flow functions f_{BL} or F_{BL}, respectively, i.e., because of (2.12) lastly on the nature of the permeability/saturation functions $k_w(S_w)$ and $k_{nw}(S_w)$. The same reasoning that at the end of the previous section can be used to infer that for a concave upward flow function $F(S_w)$ (similar to the Freundlich isotherm $n < 1$, in (2.8) a steepening of the saturation front will occur, which eventually might lead to a saturation shock (Bear, 1972; Allen, et al., 1988a). This situation is illustrated for a set of linear permeability/saturation functions as they have been proposed in oil/water displacement modeling (Coats, 1980) is illustrated in the Fig. 3.5 and will be discussed in more detail in Section 3.3.3.

3.   NUMERICAL FORWARD MODELING

3.1.   Numerical techniques for advection dominated problems

3.1.1.   Nonadaptive methods

In the present section we discuss some of the modern numerical methods which are being used for the simulation of advection-dominated transport problems. These methods can be basically grouped into the following:

(a) The **Method of Characteristics (MoC)** has long been the method of choice for solving pure hyperbolic problems such as arise in the modeling of multiphase reservoir flow as outlined above (Allen, et al. 1988a; Wheeler, 1988). To treat the effects of finite physical diffusion the MoC has been recently coupled with regular finite element methods (cf. Ewing, 1983; 1989).

(b) **Finite Difference Methods (FD)** based on either explicit or implicit time integration schemes (Lapidus and Pinder, 1982; Allen, et al., 1988b; Fletcher, 1988) and applicable to both hyperbolic/parabolic and pure hyperbolic transport problems. For the latter cases special techniques such as the **Lax-Wendroff/Friedrichs** methods and the high order **E-schemes** such as the **Godunov** and the **Rusonov** methods (Sod, 1985) have been used. **Upstream**-differentiation for the treatment of the advection term is often applied to reduce the numerical dispersion, at the expense of incorporating a finite amount of numerical diffusion into the solution.

(c) **Finite-Element (FE)** methods such as the classical **Bubnov-Galerkin** method (Fletcher, 1988) the **Collocation** technique (Allen and Pinder, 1983; Allen, et al. 1988b) and the **Petrov-Galerkin** method (Hughes, 1979; Tezduyar and Ganjo, 1986; Ewing, 1989), corresponding to the upstream finite-differencing method are starting to gain more importance in modeling of transport problems. Using the **modified equation approach** of Warming and Hyett (1974) higher order time-integration schemes such as the **Taylor Galerkin** method (Donea, et al., 1984; 1987; Baker and Kim, 1987) and the **Mass-Operator** technique (Fletcher, 1988) have been developed in recent years and appear to be more appropriate for the treatment of sharp solution fronts. An adaption of such a method which appears to reduce both the dispersion and the diffusion usually encountered in ordinary methods has been implemented in the newly developed method by Koch (1992a). We will show some of results obtained with this technique in Section 3.3

(d) **Method of Lines (MOL)** (Liskovets, 1965; Kee and Petzold, 1987) based on either a FD or collocation techniques form a special category for the solution of 1D transport problems.

Concerning the tracking and resolution of sharp fronts or shocks for pure hyperbolic systems, most of the above techniques are based on either methods of:

(1) **Front tracking**: The location of the front is computed explicitly from the **Rankine-Hugoniot entropy (jump) conditions** (Sod, 1985). While useful for hyperbolic conservation laws arising in high speed aerodynamics, they are are not suitable for the general advection-dominated transport equation (1.1), where the small amount of physical diffusion induces a finite width of the front that needs to be resolved.

(2) **Shock capturing** is based on the Lapidus approach of introducing a small amount of artificial diffusion into the PDE in order to reduce the amount of numerical dispersion in the vicinity of the front. Such are the Lax-Wendroff scheme and recently developed higher order **TVD-schemes** (Harten, 1983) but also the upwinding methods. The convergence is often proportional to the mesh width at the front. Despite their wide applications in conjunction with the FD and FE methods, shock capturing does not always work satisfactorily and is plagued by the trade-off between reducing the amount of *dispersion* on one hand and not allowing excessive smearing of the front through a too-high amount of artificial *diffusion* on the other hand (Sod, 1985; Fletcher, 1988). In fact, in most of the cases one is faced with both of these problems at the same time.

In our effort to find an appropriate method for the solution of the ADE (1) in the presence of sharp fronts, we tried several of the above FD/FE methods and the MOL available from the NAG-library. None of the methods worked satisfactorily. In Fig. 3.1 we show the results of such a simulation with a dispersion/diffusion-minimizing Taylor Galerkin FE method employing a Crank-Nicholson time-integration technique (Fletcher, 1988) and the MOL. Both the MOL and the FE solutions are contaminated by large amounts of dispersion. The NAG-MOL method in addition shows some diffusion leading to the smearing of the vertical flanks of the impulse. A more detailed investigation of the FE-method (Koch and Zhang, 1992) reveals that it produces good results for less extreme advection/diffusion ratios.

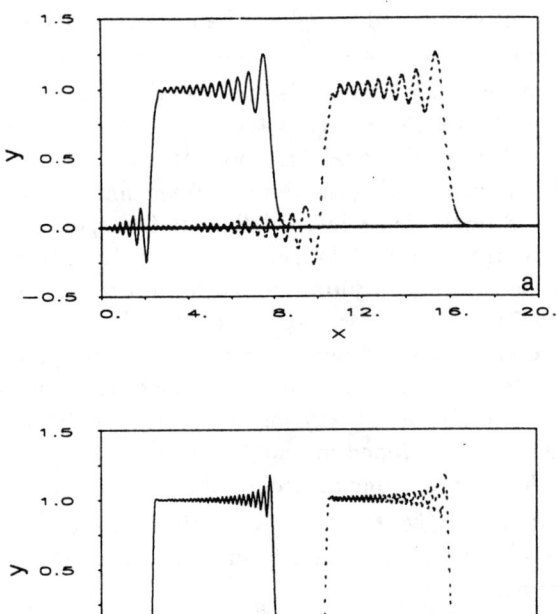

Figure 3.1. Numerical modeling of the ADE (13) with a large advection/diffusion ratio of $v/D = 10^4$ for a boxcar impulse entering the domain at $x = 0$. Shown is the impulse for two different times using a classical MOL from the NAG library (a) and a Crank-Nicholson FE method (b).

3.1.2.   Adaptive techniques

The failure of the above techniques in the presence of steep fronts is mainly related to the insufficient spatial resolution of the grid at the advancing front. To avoid an overall uniform increase of the mesh points all over the domain and to introduce redundancy of grid points in areas where the solution is not changing significantly, **adaptive methods** for pure hyperbolic and hyperbolic/parabolic transport problems have been developed in recent years. The large amount of literature in this domain can be categorized as follows:

1) **FD methods** based on mesh refinement and, possibly, time adaptation (Berger and Oliger, 1984; Petzold, 1987);

2) **FE methods**, which do, in fact, comprise the large majority of adaptive techniques (Babuska, et al., 1986; Ewing, 1986). The adaptive FE methods are can be grouped as

   - **h methods** refine the mesh in areas of large local errors computed from *a posteriori* estimates (Loehner, et al. 1984; Babuska, et al., 1986; Benner et al., 1987)

   - **moving FE methods** follow a mesh refinement along a moving front. This technique has shown to be very attractive for advecting fronts (Miller and

Miller, 1981; Miller, 1981; Adjeric and Flaherty, 1986a; 1986b; Gropp, 1987; Baines and Wathen, 1988; Johnson, et al., 1988).

- p **methods** increase the order p of the interpolation function in areas of large errors (Oden, et al., 1986);
- $h - p$ **methods** are a combination of both (Oden, et al., 1986).

3.2.   The method of lines

In the present section we will present the development of a new adaptive (transversal) method of lines (MOL) to reduce the dispersion and diffusion errors encountered in the FE-method and the classical (longitudinal) MOL as was shown in the Fig. 3.1. A detailed description and validation of the method is given in (Koch, 1992a).

3.2.1.   The longitudinal method of lines

The classical longitudinal MOL has long been known as a solution technique of a general 1D PDE's such as specified by (1.1a) (Liskovets, 1965; Kee and Petzold, 1987; Sewell, 1988). Depending on the specific approach, the idea is to discretize the spatial domain and to replace the spatial derivatives in the PDE by finite differences or a collocation expansion. Thus the PDE is transformed into a coupled system of ODE's for the $u_i(t)$ at the nodal points or for the expansion coefficients $a_k(t)$:

$$A\frac{\partial a_k(t)}{\partial t} = L\left\{f\left(t, a_k(t)\right)\right\} \tag{3.1}$$

where A is a banded matrix comprising the collocation basis functions used— usually Lagrange-interpolation functions or Hermite polynomials—and L is the spatial discretization operator. The system of ODE's is then solved by any efficient initial value ODE solver, including special features such as automatic time-stepping control (Petzold, 1987). Our numerical results, however, showed (Fig. 3.1) that the longitudinal MOL is inappropriate for the solution of advection-dominated transport problems.

3.2.2.   The adaptive transversal method of lines

In the new **transversal MOL**, unlike the longitudinal MOL which solves the coupled initial value problem (3.1) in time, a a boundary value problem (BVP) is solved in the interval a, b for each time-step using a boundary value solver. The solution is then marched in time in an implicit manner.

By rewriting Eq. (1.1a) in the form

$$\frac{\partial u}{\partial t} = u_{xx} + g(x, u, u_x) \tag{3.2}$$

and performing a Euler backward-time discretization of the left side of Eq. (3.2) one gets for the time-step $(n + 1)$:

$$\frac{(u^{n+1} - u^n)}{\Delta t} = u_{xx}^{n+1} + g\left(x, u^{n+1}, u_x^{n+1}\right) + O(\Delta t) \tag{3.3}$$

which shows that the technique is first-order accurate in time.

Reordering Eq. (3.3) gives

$$u_{xx}^{n+1} = -g(x, u^{n+1}, u_x^{n+1}) - \frac{u^{n+1}}{\Delta t} + q^n \qquad (3.4)$$

where $q^n = -u^n/\Delta t$ is the scaled solution at time-step (n). Eq. (3.4) is a second-order BVP.

For the solution of the first-order hyperbolic PDE (2.3) the same procedure results into a first-order BVP:

$$u_x^{n+1} = -g(u, x) \left\{ \frac{(u^{n+1} - u^n)}{\Delta t} \right\} \qquad (3.5)$$

where $g = (f_{fl}(u, x))^{-1}$, the inverse of the flux function.

The ODE's (3.4) and (3.5) can be solved for the time-step $(n + 1)$ by any two-point BVP solver. Starting with an initial solution u^0 the solution is then marched in time by inserting for each time step the previous solution u^n into Eqs. (3.4) or (3.5). In account of the backward Euler time-integration scheme the transversal MOL is unconditionally stable. However, we will see that in spite of this desirable property the first-order time integration scheme (3.3) requires small time-steps Δt has to get sufficient accuracy and to avoid numerical diffusion.

For the numerical solution of the BVP problem we have chosen the powerful boundary value solver COLSYS (Ascher, et al., 1981; Bader and Ascher, 1982; Ascher, et al., 1988) which solves a system of general boundary value problems of the form

$$u_i^{(m_i)} = f_i(x, z(u)) \qquad 1 \le i \le d \qquad a \le x \le b \qquad (3.6)$$

and general BC's

$$g_j(z(y(\zeta_j))) = 0 \quad 1 \le j \le m^* \qquad (3.7)$$

where

$$u(x) = (u_1(x), \ldots, u_d(x))^T,$$

$$m^* = \sum_{j=1}^{d} m_i,$$

$$a = \zeta_1 < \zeta_2 < \ldots \zeta_{m^*} = b,$$

and

$$z(u(x)) = (u_1(x), \ldots, u_1^{(m_1-1)}(x), \ldots, u_d(x), \ldots, u_d^{(m_d-1)}(x))^T$$

and f and g are generally nonlinear functions. The points ζ_i are general fix points of the solution, which for the present application are just the two boundary points a, b. The number of ODE's reduces to $d = 1$ here, which is in contrast to the longitudinal MOL (3.1) which requires the solution of a system of N coupled ODE's.

The COLSYS method is based on a collocation technique using Lagrange's interpolation functions of order up to $k = 7$ evaluated at the Gaussian points in the subinterval $x_i, x_{(i+1)}, i = 1, N$, where $N + 1$ is the total number of grid points in the model interval a, b.

An error estimate (Babuska and Rheinboldt, 1978) shows that the norm of the spatial error in the solution is given by

$$\|u(x) - u_\pi(x)\| = O(h^{k+m})\|u^{(k+m)}\| \tag{3.8}$$

where $u_\pi(x)$ is the numerical solution obtained on the mesh, π, with mesh size h and m is the order of the BVP (i.e., $m = 2$ for the parabolic Eq. (3.4) and $m = 1$ for the hyperbolic Eq. (3.5). Our numerical experiments showed the optimal efficiency of the code for $k = 6$, so that Eq. (3.4) results in a very high spatial accuracy.

Another desirable property of the COLSYS code is the availability of an adaptive mesh generator which is used to adapt the mesh in regions where the spatial solution is changing rapidly. Using an error estimate of the solution similar to Eq. (3.8) regions of high spatial errors for a particular mesh can be identified. By using a monitor function τ (Ascher, et al., 1988) the mesh is then redistributed and/or the number of grid points is increased in the critical regions in order to equilibrate the errors and to reduce them beneath a prescribed tolerance limit TOL.

Although numerical experiments by Ascher, et al. (1988) show this adaptation technique to be very powerful even for sharp transition layers in the solution $u(x)$ of the BVP, for use in the transversal MOL a number of technical changes had to be made to avoid 'runaway' of the adapting mesh as the advection front advances with time (see Koch, 1992, for details).

From Eq. (3.8) and the mesh adaptation it follows that an extremely high spatial accuracy of the present transversal MOL can be achieved. However this asset of the MOL, which also holds for any other adaptive technique discussed in the previous section, may be lost to a certain extent in the solution of a time-dependent PDE. In fact, the overall accuracy is determined by both the spatial and the time-discretization error. From Eq. (3.3), the latter is seen to be of first-order which means that the overall accuracy of the present MOL can only be obtained for very small time-steps Δt. In fact, our numerical simulations in the following section do reveal some amount of numerical diffusion of the first-order MOL for insufficiently small time-steps Δt.

Donea, et al. (1984; 1987) were the first to recognize the frequent imbalance of the spatial and temporal discretizations as the source for excessive numerical dispersion and/or diffusion in regular FD and FE approximations of hyperbolic systems. Using higher order terms in the Taylor expansion for the time derivative, they derived a series of **Taylor Galerkin methods** which do appear to improve the dispersion/diffusion properties of classical schemes such as the Lax-Wendroff and the Crank-Nicholson methods. Baker and Kim (1987) have generalized the Donea approach by developing the **Taylor Weak Statement (TWS)** technique which is able to reproduce most of the well known FD and FE techniques.

Following these ideas we have implemented for the pure hyperbolic equation (1.2c) whose first-order MOL discretization was given by Eq. (3.5) a **third-order Taylor time-discretization technique** into the transversal MOL (Koch, 1992a). As shown by some of the results in the next section, the Taylor MOL improves the accuracy of the solution and, compared with the first-order MOL technique, for the same time step Δt, is able to significantly reduce the numerical diffusion . However, the Taylor MOL appears to be less robust than the first-order MOL technique, in particular for strong nonlinear advection flow problems as described in Section 2 (Koch, 1992a).

3.3.   Simulation examples

In the following section we apply the new adaptive MOL to some of the special forms of the ADE described in Sections 1 and 2, governing

(a) pure linear advection (Eq. 1.2a);

(b) linear advection, linear adsorption and diffusion (Eq. 2.9);

(c) nonlinear hyberbolic plug-flow adsorption (Eq. 2.7);

(d) the nonlinear hyberbolic two-phase Buckley-Leverett flow (Eq. 2.14).

The application of the MOL to the general nonlinear full hyberbolic/parabolic ADE Eq. (2.4) for adsorption including diffusion and also to the nonlinear full hyberbolic/parabolic two-phase Buckley-Leverett equation (2.10), including capillary effects, is in progress.

3.3.1.   Linear advection

By applying the first-order MOL-formalism to the linear hyperbolic advection equation (1.2a), one obtains the fist-order BVP as stated in Eq. (3.5). The Taylor MOL, on the other hand, results in a second-order BVP (see Koch, 1992a for details).

Figs. 3.2 and 3.3 show the results for the first-order method and the Taylor Mol method, respectively, of the pure advection of a front with a front width of $\epsilon = 0.01$ entering the domain at $x = 0$. Figs. 3.4 and 3.5 show the same situation for an advected sinusoidal pulse. The solutions are presented for different values of the Courant-number CFL and of the variable TOL. Note that because of the varying adaptive mesh size, the usual definition of the CFL number, (i.e., CFL $= v\Delta t/\Delta x$ where Δt is the time-step and Δx the (fixed) mesh size) is not appropriate here and Δx is replaced here by the effective width ϵ of the front or the sinusoidal pulse, respectively. The variable TOL is an input variable for the COLSYS BVP solver and steers the amount of adaptation and mesh refinements for each time-step such that the *a posteriori* estimated spatial error $e(x)$ in the solution as given by Eq. (3.8) is less then the requested tolerance TOL. However, as outlined in the previous section, this spatial error estimate $e(x)$ reflects only the solution for one particular time-step, but is not able to predict the error propagation of the solution with time. The latter is determined by the chosen time discretization. This difficulty with estimating mathematically the exact total error has to be considered as one of the major pitfalls of adaptive techniques for time-dependent PDE's.

The error-effects can be clearly seen in Figs. 3.2–3.5 when comparing the solutions for the first-order MOL and the Taylor MOL. For the same Courant number CFL the Taylor MOL is basically diffusion-free while the first-order MOL, despite the smaller requested spatial error tolerances, TOL, still reveals a large amount of diffusion but practically no dispersion. It means that the higher spatial precision in the first-order MOL cannot make up for the low time discretization of this method. The Taylor method with a first-order time-derivative approximation on the other hand is the other hand produces much better results for both smaller time-steps and less stringent spatial precision. For the first-order method is direct positive correlation between the Courant number and the amount of numerical diffusion generated. By using very small time-steps the diffusion of this technique can be reduced to an arbitrary small amount.

To investigate the effects of the mesh adaptation and its need for obtaining a reliable solution at all a series of simulations with an a *a priori* fixed equidistant

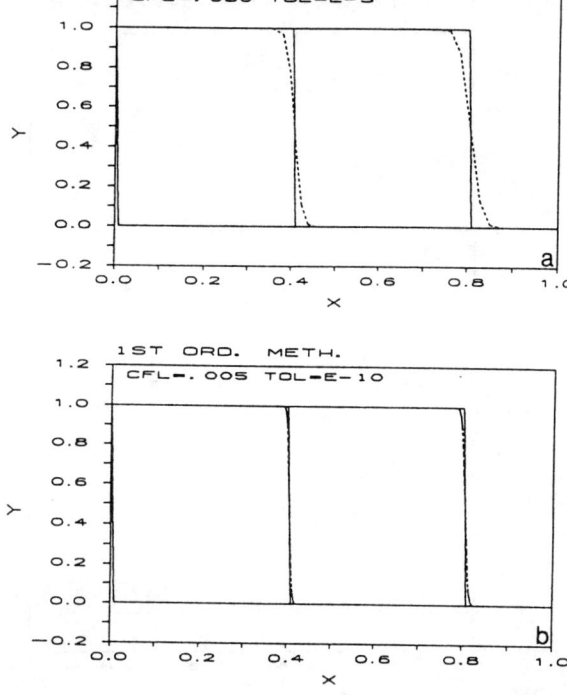

Figure 3.2.   Pure advection of a front for the first-order MOL for different values of the Courant number CLF and the spatial error tolerance parameter TOL.

grid was performed (Koch, 1992a). The results of these investigations showed that, whereas the first-order MOL still exhibited a large amount of numerical diffusion, for the same Courant number the Taylor MOL on the other hand advected the impulse perfectly well. And this, despite the fact that the *a posteriori* estimated spatial errors for the Taylor OL where generally some orders of magnitude higher than the ones obtained with the first-order MOL. These results corroborated the fact that the spatial error does not by itself indicate much about the overall reliability of the MOL—or of any spatial adaptive method—for the solution of the PDE. These, rather surprising results do point out the need for both, high temporal and spatial discretizations in the numerical PDE-solver.

In terms of the relative computational efficiency of both methods our results appear to show a large advantage for the Taylor MOL, which for the same amount of numerical diffusion as obtained with the first-order MOL, can use time-steps Δt about 10 times as large as are needed in the first-order MOL. However, while the first-order MOL is only diffusive but practically dispersion-free, the Taylor MOL reveals for non-optimal Courant numbers and non-optimal spatial error tolerances some amount of dispersion (Koch, 1992a). Thus, due to the additional antidiffusion term from the second-order spatial derivative in the BVP formulation of this method some of the excessive diffusion of the first-order MOL has been traded for dispersion that might lead to a divergence and eventual complete breakdown of the numerical solution

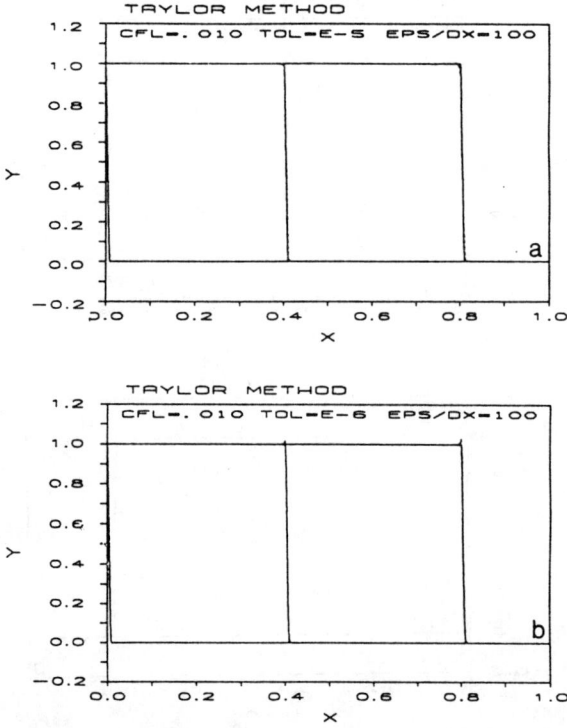

Figure 3.3. Same as Figure 3.2 using the Taylor method. $\epsilon/\Delta x$ is the ratio of front width to the finite-difference increment used in the Taylor MOL (Koch, 1992a)

of the Taylor MOL. We have found generally, that despite its inherent diffusion (which can be monotonically reduced by employing small time-steps), the first-order MOL technique is much more robust than the Taylor method, particularly for the solution of the nonlinear hyperbolic plug-flow and Buckley-Leverett models discussed in Sections 3.3.3 and 3.3.4.

3.3.2.   Linear advection with diffusion

The application of the first-order MOL formalism to the linear ADE (2.9) results in a second-order BVP as stated in Eq. (3.4). Although a higher order Taylor approximation for the time-step in the general ADE has been shown to be possible (Donea, et al., 1984), we are still investigating such an implementation into the transversal MOL.

The Fig. 3.6 shows the numerical results for two advected/diffused rectangular pulses obtained with the adaptive first-order MOL. These models are similar to the one of Fig. 3.1 for which the longitudinal MOL and the FE method failed. The only difference in the two models is the ratio of the v/D which is 10^9 and 10^3, respectively. The time-steps for both cases are $\Delta t = 10^{-3}$. The pulses are shown for 4 different times. The finite duration of the pulse has been modeled through a time-dependent boundary condition.

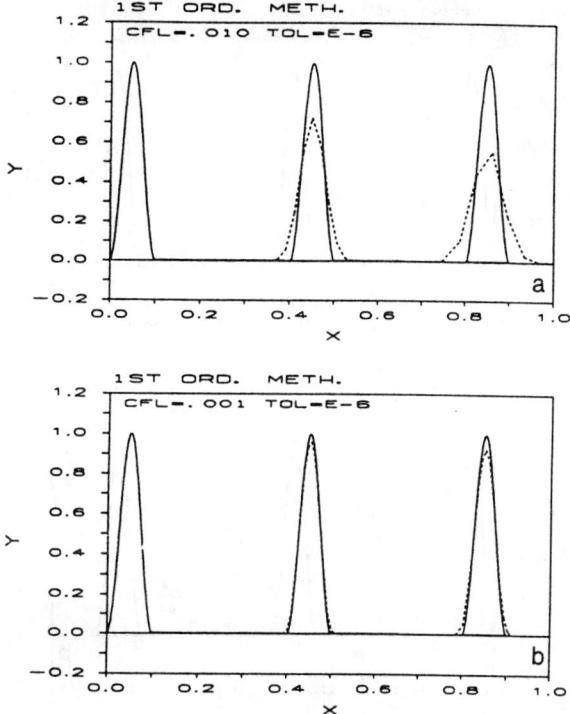

Figure 3.4. Same as Fig. 3.2 but for a sinusoidal pulse. The Courant number is based in this case on the base width $\epsilon = 0.1$ of the pulse.

While the model with the large value of v/D corresponds to a nearly pure hyperbolic situation and produces no relevant amount of physical diffusion, the pulses of the second model are in fact diffusing with time. With the very small time-steps for both models being equal, numerical diffusion does not play any significant role here. Nevertheless it is recognizable at the flanks of the pulses of the left model in Fig. 3.6. By reducing the time-step of the marching process further, we were able to decrease the numerical diffusion beneath the optical detection limit. Once again we have found that the accuracy of the method depends more on the size of the time-step than on the spatial mesh precision. Because the total CPU-time is proportional to the number of time-steps used, the development of a higher order Taylor technique appears to be the more indispensable.

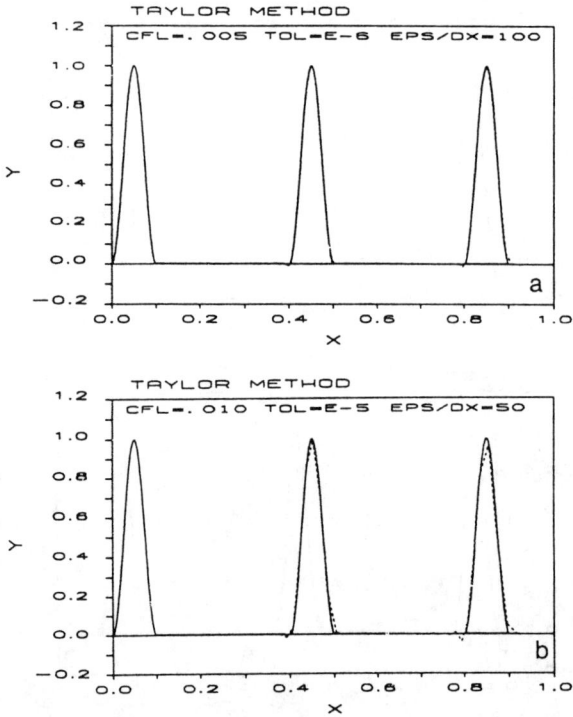

Figure 3.5. Same as Fig. 3.4, using the Taylor method.

3.3.3. Nonlinear advective adsorption

In the following two sections we will discuss the application of the transversal MOL to the nonlinear hyberbolic plug-flow adsorption model (Eq. 2.7). Although there is a provision in the COLSYS code to solve a nonlinear BVP by either a damped Newton linearization or a Jacobi iteration technique, some particular strategies had to be developed for the application of this BVP-solver in the MOL solution of the general time-dependent PDE (1.1c) or (1.2c) (Koch, 1992) in order to avoid divergence of the iteration process with ongoing time when the fronts are steepening.

Fig. 3.7 shows the results of the simulation for the two different kinds of Freundlich's isotherms (2.8), namely one which is concave upwards ($n = 0.5$) and the other which is concave downwards ($n = 2$). In both cases, an increasing skewing of the original bell-shaped pulses with time can be observed. Compared with the propagation of the pulse with linear isotherm with $n = 1$ (not shown here), the pulses for the two nonlinear isotherms with $n = 2$ and $n = 0.5$ are retarded or advanced, respectively.

Fig. 3.7 is basically a quantitative confirmation of the arguments made at the end of Section 2 and reveals the capacity of the transversal MOL to capture evolving fronts. Note that no physical diffusion has been incorporated in this model. This would lead to a mitigated front sharpness. The application of the MOL to the nonlinear full hyberbolic/parabolic ADE (2.4) using various adsorption isotherms and including physical diffusion is in progress and will be contrasted with the work of Gupta and Greenkorn (1973) who used a classical FD method to model such a process

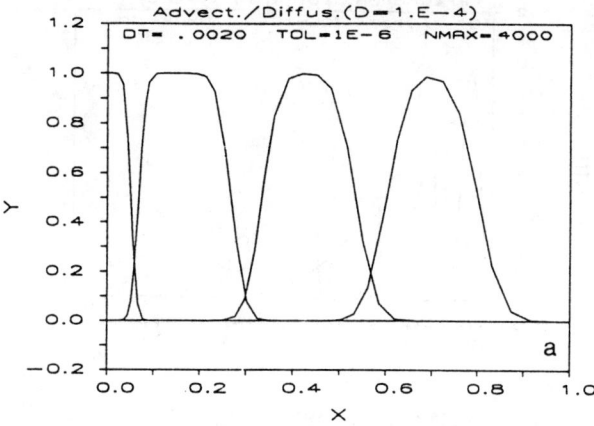

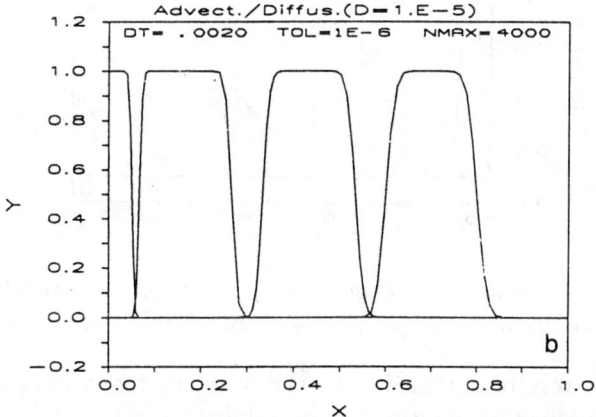

Figure 3.6. Linear advection/diffusion of a pulse for two different values of the diffusion coefficient D. The advection velocity for both models is v = 0.1. Only the front of the first pulse on the left can be seen.

in an infinite half-space. The same technique was used by Satter, et al., (1980) for a finite column length. In view of the mentioned numerical problems, which the classical FD methods demonstrated in the presence of sharp fronts, the results of these authors need to be re-evaluated.

3.3.4. Nonlinear advective Buckley-Leverett flow

As a final example we show the application of the transversal MOL to the nonlinear hyberbolic Buckley-Leverett flow equation (2.14). As outlined in Section 2.2. the strong tendency of the BL-equation to form sharp fronts and finally to develop shocks makes it a particularly challenging test case for new numerical methods. Techniques such as the **Random Choice method** (Concus and Proskurowski, 1979), a classical FE-collocation method (Allen and Pinder, 1983) or Moving FE (see Section 3.1) have been tested on the BL-problem. Except for the

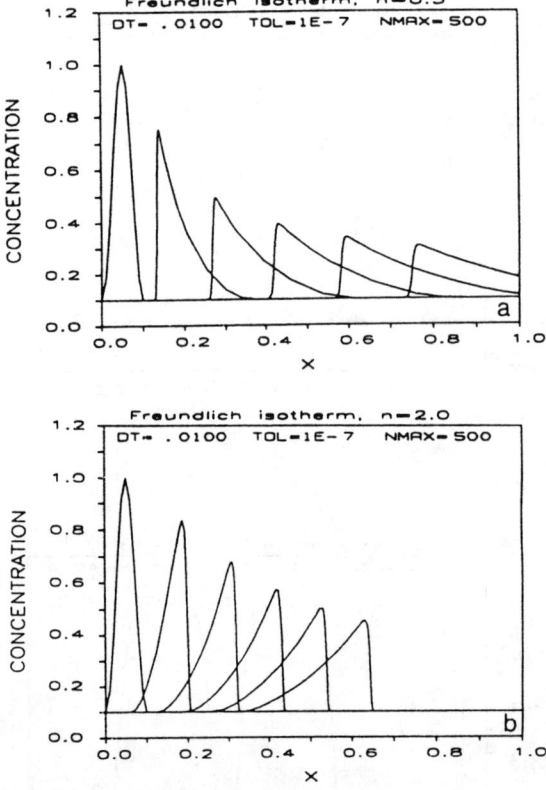

Figure 3.7. Simulation of the adsorption plug-flow model (2.7) for two different exponents n for the Freundlich's isotherm (2.8).

last technique, the results obtained with the other methods have not always been satisfactory and do show large dispersion and diffusion in the vicinity of the front.

In the solution of the BL equation we have used for the nonlinear fractional flow-function $F_{BL}(S_w)$ (2.12) the following standard expression.

$$F_{BL}(S_w) = \frac{S_w^2}{S_w^2 + \alpha(1 - S_w)^2} \tag{3.9}$$

where, in agreement with previous publications, $\alpha = 0.5$ has been taken. From Eq. (2.12) it can be seen that Eq. (3.8–3.9) results from a quadratic permeability/saturation relationship for the wetting and nonwetting phases k_w and k_{nw}, respectively. The corresponding diagrams are shown in the Fig. 3.6 and are representative of a water/oil two-phase system (Coats, 1980) or for an air/water system in the partially saturated groundwater zone (Van Genuchten, 1980).

As expected from the shape of the flow function F_{BL} which has an inflection point (i.e., the derivative of F_{BL} has a maximum) a saturation shock occurs. Using the Rankine-Hugoniot jump conditions for the propagation speed of the shock and the **Welge tangent construction** (Allen, et al., 1988a) the analytical solution of the problem can be obtained in an approximate manner. Our numerical results are consistent with the analytical solution and appear to be superior to the numerical solutions obtained by the authors above, in the sense that the shock is better resolved. Since our solutions are very close to the ones obtained by the moving FE-technique

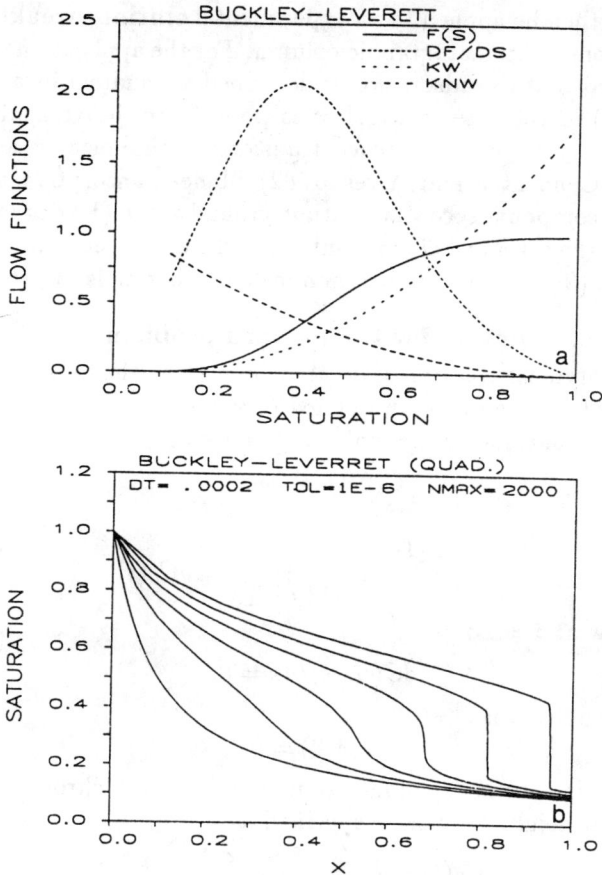

Figure 3.8. Simulation of Buckley-Leverret flow using the quadratic flow function F_{BL} (3.9). (a) Shown are the permeability/saturation function k_w and k_{nw} for the wetting and and nonwetting phases, respectively; the flow function F and its derivative. (b) Saturation profiles for different times.

(Section 3.3), the advantage of automatic mesh adaptation for solving this problem becomes obvious.

It turns out that the form and propagation of the saturation profile is extremely dependent on the nature of the fractional flow function F_{BL}. Thus, the determination of the exact quantitative functional relationship of $k_w(S_w)$ and $k_{nw}(S_{nw})$ in Eq.(2.12) between permeability and saturation for various two-phase systems, such as oil/water in reservoir modeling or air/water for the unsaturated soil zone, has been a major topic of current research (Van Genuchten 1980; Luckner et. al, 1989). The effects of the various permeability/ saturations function on the movement of the saturation front in the unsaturated zone using the present MOL has been modeled in detail by (Koch, 1992).

4. INVERSE MODELING OF AN ANALYTICAL MODEL

In the following sections we will present the development of an inverse technique to determine directly a few major physical transport parameters from a known analytical solution of the ADE (2.9), i.e. after forward modeling has been performed.

The technique will then be applied to an experimental **elution-breakthrough curve** (BTC) obtained from a chromatographic column. For the application of the analytical model, it is assumed that the adsorption is described adequately by a linear isotherm, i.e., $n = 1$ in (2.8) in the case in which it is possible to derive analytical solutions for the ADE (2.9) by means of Laplace transform techniques or series expansions (Bear, 1979; Van Genuchten and Alves, 1982). Linear adsorption appears to hold for many organic compounds contaminating groundwater (Karickhoff, 1981). The implementation of the inverse technique into the efficient numerical MOL to solve the more general ADE (2.4) including such nonlinear isotherms is in progress.

4.1.   An analytical solution for the forward problem

We present a typical solution $C(x,t) = u(x,t)$ of the ADE (2.9) with linear adsorption in an infinite half-column defined in $x > 0$.

The boundary conditions at the inlet $x = 0$ are of either Dirichlet's type

$$C(0,t) = C_0(t) \tag{4.1}$$

or of Cauchy's type

$$\left. \left(-D\frac{\partial C}{\partial x} + vC(t) \right) \right|_{x=0} = vC_0(t) \tag{4.2}$$

and for the outflow at $x = \infty$

$$dC(\infty,t)/dx = 0. \tag{4.3}$$

while the initial condition is

$$C(x,0) = 0. \tag{4.4}$$

In (4.1) $C_0(t)$ is a time-dependent source which for a chromatographic elution pulse with constant height C_0 can be specified as

$$C_0(t) = \begin{cases} C_0 & \text{for } 0 < t < t_0 \\ 0 & \text{for } t > t_0 \end{cases} \tag{4.5}$$

The analytical solution is then given by (Van Genuchten and Alves, 1982)

$$C(x,t) = \begin{cases} C_0 A(x,t) & \text{for } 0 < t < t_0 \\ C_0 A(x,t) - C_0 A(x,t-t_0) & \text{for } t > t_0 \end{cases} \tag{4.6a}$$

where

$$A(x,t) = \frac{1}{2}erfc\left[\frac{(Rx - vt)}{2(DRt)^{1/2}}\right] + \left[\frac{v^2t}{\pi DR}\right]^{1/2} \exp\left(-\frac{(Rx - vt)^2}{4DRt}\right)$$

$$- \frac{1}{2}\left[1 + \frac{vx}{RD} + \frac{v^2t}{RD}\right] \exp\left(\frac{vx}{D}\right) \tag{4.6b}$$

$$erfc\left[\frac{(Rx + vt)}{2(DRt)^{1/2}}\right]$$

This expression, which encompasses also the classical BTC when t_0 is very large, has been evaluated numerically for various combinations of v, R and D. Special consideration has to be taken for the simulation of sharp fronts which correspond to large ratios of v/D in Eq. (4.6) to avoid overflow and rounding errors in the second term in Eq. (4.6). In fact, for $v/D \to \infty$, the ADE will be asymptotically hyperbolic and the solution (4.6) breaks down. Analytical solutions of the form (4.6) are being used frequently for the modeling of the migration of contaminants in one-dimensional groundwater flow (Javandel, et al. 1985; Koch and Zhang, 1989).

4.2.   Parameter Estimation

The free parameters determining the solution (4.6) are the transport coefficients v, R and D, and t_0 given from the initial condition, respectively. C_0 can be eliminated right away by employing the normalized solution C/C_0 in (4.6). However, in the present case of linear adsorption it follows directly from the ADE (2.9) or the analytical solution (4.6) that v, D and R cannot be determined independently, i.e. their inverse solution is not unique. Thus, it is only possible to determine the ratio of v/R in Eq. (2.9) so that v needs to be fixed to get a unique solution. In the case of solute transport in groundwater flow, v can be determined from the piezometric head distribution using Darcy's law (Bear, 1972), while for the chromatographic experiment, v is directly related to the injection flow rate. t_0 in Eq. (4.5) is the inlet width of the injection pulse and determines the total amount of solute entering the system.

If the physical model, as represented by the advection-dispersion equation, were known with absolute certainty to be correct, and observations of the concentration could be made without error, a deterministic inversion model would be an appropriate choice. The assumptions of a correct physical model and perfect observations are usually inappropriate for practical transport problems. Thus, the chosen model may contain errors in the model as well as in the observations. The inversion model may be represented as:

$$C_{\text{obs}} = C_{\text{mod}} + E_R + E_s, \qquad (4.7)$$

where C_{obs} is the observed concentration, C_{mod} is the deterministic prediction from the physical model, E_R is the random error, and E_s is the systematic error due to an incorrect physical model. C_{mod} depends on the parameters and the boundary and intial conditions chosen for the model (Knopman and Voss, 1987).

Inverse modeling can be viewed as solving a nonlinear multiparameter estimation problem through a time history matching of the observed elution profile $C(x_0, t)$ at a position x_0 by the theoretical model (4.6).

This problem can be cast into a general nonlinear optimization or minimization problem of the objective function F, i.e., we search for the model vector $X = (D, R, t_0)^T$ such that

$$F = \|y - f(X)\|^2 \longrightarrow \quad \text{minimum} \qquad (4.8)$$

where $y = (C_1, C_2, \ldots, C_n)^T$ is the vector of observed concentrations C_i taken at time t_i and $f(X) = C(x_0, t)/C_0$ is described by the theoretical model (4.6).

The most widely used techniques for the solution of Eq. (4.8) are various members of the family of **gradient techniques** (Fletcher, 1980; Gill, et al., 1981) which in the ith iteration step search for the minimum of Eq. (4.8) in the direction of the gradient g_i of the objective function F_i evaluated for $X = X_i$

$$X_{i+1} = X_i + \lambda_i S_i g_i \qquad (4.9)$$

with $g_i = \nabla F_i$, $S_i = H_i^{-1}$ is an approximation of the inverse of the Hessian, H_i, of F_i, defined by $H_i = \nabla^2 F_i$, and λ_i is a step-length reduction parameter chosen to ensure a reduction of the objective function F_i.

Depending on the choice of the matrix S_i (4.9) reduces to different technique such as:

a) **Method of steepest descent (SD)**: $S_i = I$, the identity matrix,
b) **Newton or quasi-Newton methods**: $S_i = H_i^{-1}$.
  For the objective function, F, (4.8) one obtains for the Hessian, H_i, the expression

$$H_i = 2(\nabla f_i)^T \nabla f_i + 2(\nabla f_i)^2 r_i, \tag{4.10}$$

where r is the residual $r_i = y_i - f_i(X_i, t)$, and for the gradient $g_i = \nabla F_i$

$$g_i = -2(\nabla f_i)^T r_i. \tag{4.11}$$

The well known **Gauss-Newton (GN) method** results by dropping the second term in Eq. (4.10) and thus approximating H_i by

$$H_i = 2\nabla f_i^T \nabla f_i, \tag{4.12}$$

and substituting this and (4.11) into (4.9) and setting $\lambda_i = 1$:

$$X_{i+1} = X_i + (J_i^T J_i)^{-1} J_i^T r_i, \tag{4.13}$$

where $J_i = \nabla f_i$ is the **Jacobian** of the model function.
  The Gauss-Newton minimization technique (4.13) often fails in cases where the objective function F is degenerate and exhibits a badly defined local minima. Then the Jacobian J_i is ill-conditioned and overshooting beyond the minimum of the function F occurs. The condition of the Jacobian J can be improved by adding a damping parameter, k, on the diagonals of J. Thus one obtains the **Levenberg-Marquardt (LM) technique** (Levenberg, 1942; Marquardt, 1942)

$$X_{i+1} = X_i + (J_i^T J_i + kI)^{-1} J_i^T r_i, \tag{4.14}$$

which has been applied with success to many nonlinear inversion problems in geophysical sciences (cf. Koch, 1991).
  Depending on the damping parameter k, the LM method encompasses both the GN method and the method of steepest descent. In fact, for $k \to 0$ the GN method is obtained. while for $k \to \infty$. the LM method behaves like the SD method. From the general convergence properties of these two methods, the strategy for choosing the optimal k-parameter can be developed. While the GN method converges quadratically near the minimum of the objective function F, the SD method behaves in a more stable manner far away from the minimum. It follows that k should be chosen large at the beginning of the search and relaxed as the search vector approaches the minimum. Following the main ideas of the original algorithm of Marquardt (1963), we have implemented such a search strategy. One major prerequisite of the appropriate strategy is that the step vector delta $\Delta X_i = X_{i+1} - X_i$ is adequate for each iteration step which means that the residual sum of squares $F_i = r_i^T r_i$ decreases from one iteration to the next, i.e. $F(x^{i+1}) < F(x^i)$. The iterations are terminated when $F(x^i) < \epsilon$ or $|\Delta x|^2 / |x|^2 < \delta$ where ϵ and δ are prescribed bounds.
  Note that for both the GN and LM methods, the second derivative in the Hessian (4.11) has been neglected. While this might not be problem in the vicinity of the minimum of F, where the residual r_i is small, it can impede the convergence of the procedure when starting the searching process far off the minimum.

The second method we have used is a **Variable Metric method** (Fletcher, 1980; Gill, et al., 1981) which starts also from (4.10) but uses an updated approximation

$$S_{i+1} = S_i + dS_i \qquad (4.15)$$

for the inverse Hessian H_{i+1}^{-1}.

The initial inverse Hessian approximation S_1 is usually taken as the identity matrix I. Various forms for the update dS_i of S_i have been proposed and have resulted in the special formulae of Davidon; Davidon, Fletcher, Powell (DFP) or of Broyden, Fletcher, Goldfarb and Shanno (BFGS), respectively (Fletcher, 1980). For the present study we have implemented the **BFGS** variable metric method (Shanno and Phua, 1980) which up to date appears to be the most stable and widely used of the three different techniques, particularly for large-scale minimization problems (cf. Navon and Legler, 1987).

4.3.   Inversion of a chromatogram.

Both the BFGS-technique and the LM-method have been applied to the inversion of an experimental chromatogram of a relatively inert chemical compound. In contrast to natural transport processes of a solute in the field, chromatographic systems are reasonably well defined, despite lasting uncertainties in the proper choice of the boundary conditions and the exact nature of the physical adsorption process. In fact, for nonlinear isotherms, the determination of the latter has to be reserved for the inversion using the numerical MOL of Section 3 for the solution of the forward problem.

Two sets of numerical experiments have been performed. In the first one we assume that the total vector $E = E_R + E_s$ in Eq. (4.7) is essentially made of the systematic error E_s; i.e., the measurement errors are negligible. In the second test, in addition normally distributed observational errors are added on to the observed chromatogram.

The first method we use here is the BFGS-method. The following values for the three estimation parameters were obtained: $D = 0.117, R = 1.021$ and $T_0 = 10.091$, respectively. As one can see from Fig. 4.1, the theoretical model is able to fit the observed chromatogram very satisfactorily. The *a posteriori* minimum residual sums of squares χ^2 which is equal to the minimized function F in Eq. (4.1) is $\chi^2 = 0.185$ and was obtained after some 30 iterations. Its nonzero value hints to errors E_R in the data, but more to an inappropriate model, either due to nonlinear adsorption effects, an inadequate analytical solution which assumes an infinitely long column, or due to inappropriate boundary conditions. A future goal of the investigations will be a more detailed statistical analysis of the parameter space and to apply techniques of model discrimination (Beck and Arnold, 1977, Knopman and Voss, 1987) to attempt to separate some of these effects.

Fig. 4.1 shows also the results with normally distributed random error added to the original data. The BFGS method behaved very stable and was able to reduce the residual sums of squares to $\chi^2 = 0.621$ while the estimated parameters values obtained are $D = 0.114, R = 1.020, T_0 = 10.078$, respectively, i.e., values very close to the error-free case above.

The second method, we apply to the same data set, is the LM-method as described above. Compared with the monotonous convergence of the BFGS-method,

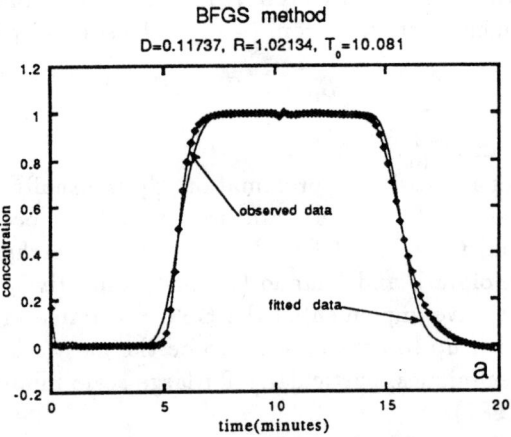

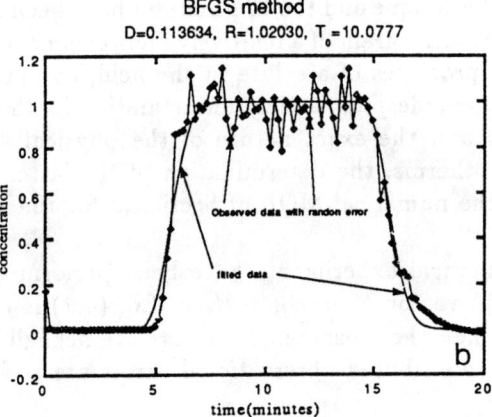

Figure 4.1. Results of the inversion using the BFGS method. a).
No additional errors are included into the chromatogram. b).
Addition of normally distributed errors into the chromatogram.

the corrected parameter vector of the LM-method gets trapped after some 10 iterations in a local minimum and the LM-technique is not able to reduce the residual beneath the final value of $\chi^2 = 1.52$, which is an order of magnitude higher than the one obtained with the BFGS-method. The corresponding solution parameters are $D = 0.142, R = 1.1628, T_0 = 9.431$, respectively. These values result in some misfit of the observed data (Fig. 4.2a).

The solution for the LM-technique including random errors of the same variance as above is shown in Fig. 4.2b. We obtain a $\chi^2 = 2.40$, while the parameter values are D=0.142, R=1.162 and T_0=9.431, respectively; i.e,. they are identical to the ones calculated for the case without statistical errors. This interesting behaviour of the LM-technique in finding the same location of the local minimum regardless of the errors in the data is certainly a consequence of the construction of the Hessian-

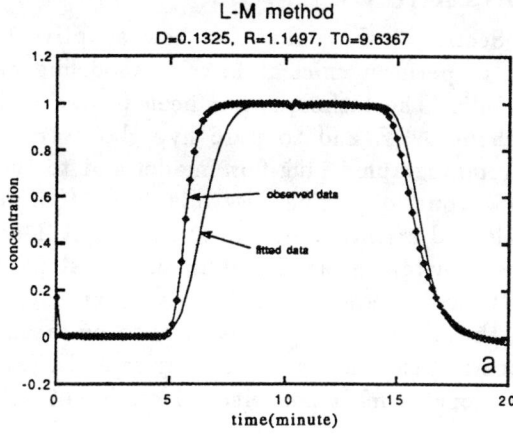

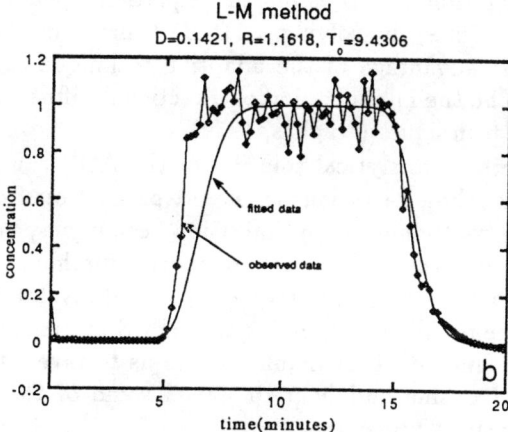

Figure 4.2. Same as Fig. 4.1 but using the LM method. a): No random errors included. b): Addition of normally distributed errors.

matrix, H_i (4.12) which does neglect the second term in Eq. (4.10) containing residual information on the Gauss-Newton step-vector ΔX_i.

The above results show clearly that the BFGS-variable metric method is superior to the Levenberg-Marquardt technique. From the computational point of view it is also more efficient since it avoids the inversion of the Jacobian matrices as required for all Gauss-Newton methods. Moreover, the variable metric method proves to be well-suited for a vector-computer, due to the fact that the updated Hessian, H_{i+1}, is evaluated by performing a series of scalar-products of vectors which can be vectorized very efficiently. Thus, the BFGS-method will be our method of choice for the inversion of nonlinear transport-parameters using the numerical forward modeling techniques of Section 3.

5. CONCLUSIONS AND OUTLOOK

The results in Section 3 show that the new adaptive transversal method of lines (MOL) is able to perform efficient forward modeling of the 1D-ADE in the presence of sharp fronts. The technique has been proven to be applicable to both the parabolic/hyperbolic ADE and to pure hyperbolic transport equations, such as the nonlinear chromatographic plug-flow model and to the nonlinear two-phase Buckley-Leverett flow equation. The new adaptive MOL-method appears to be unconditionally stable and essentially dispersion-free, but does show some diffusion, which can be minimized by taking sufficiently small time-steps. The inherent diffusion of the first-order MOL can be reduced significantly by employing a second-order time-stepping procedure, though more provisions have to be taken to prevent numerical dispersion. The numerical superiority of this Taylor MOL reveals clearly that mere space-adaptation, as done in most adaptive solvers is not sufficient for an efficient solution of a PDE.

For the future we see the possibility of using the present adaptive MOL-technique for the solution of 2D transport equations, via a **splitting-operator technique** (Fletcher, 1988), by solving a 1D BVP problem alternatively along each coordinate axis. The advantage of the adaptive technique over classical fixed grid methods, appears to be the minimization of directional diffusion (Ewing, 1986), often encountered in such techniques (Hughes, 1979).

In the case where an analytical solution of the ADE can be obtained, we have performed inverse modeling for various transport parameters from a chromatographic elution profile using two nonlinear optimization techniques, namely the Levenberg-Marquardt method and the BFGS-variable metric method. The latter technique proved to converge better than the first one, particularly for noisy data. Presently we are working on the problem to combine the BFGS-inverse method with the MOL-forward modeling technique. This should enable us to invert for nonlinear diffusion and adsorption mechanisms and help to unveil some of the physical mechanisms involved in such transport processes.

Acknowledgements

We would like to thank Dr. W.T. Cooper from the FSU Department of Chemistry for providing us with the chromatographic data. The present work has been sponsored by the Department of Energy through a Contract No. DE-FC05-85ER250000. Additional support has been provided by the Florida State University through the allocation of supercomputer resources on the ETA10.

REFERENCES

Adjeric, S. and J. E. Flaherty, A moving finite element method with error estimation and refinement of one-dimensional time dependent partial differential equations, SIAM J. Numer. Anal, 23, 778–796, 1986a.

Adjeric, S. and J. E. Flaherty, A moving mesh finite element method with with local refinement for parabolic partial differential equations, Comp. Meths. Appl. Mech. Eng. 56, 3–26, 1986b.

Allen, M. B. and G.G. Pinder, Collocation simulation of multiphase porous-medium flow, Soc. Pet. Eng. J., 23, 135-142, 1983.

Allen M.B., G. A. Behie and J. A. Trangenstein, Multiphase flow in porous media, Mechanics, Mathematics and Numerics, Springer Verlag, Berlin, New York, 1988a.

Allen, M.B., I. Herrera and G. Pinder, Numerical modeling in science and engineering, John Wiley & Sons, New York Chichester, 1988b.

Ascher, U. M., J. Christiansen, and R.D. Russel, Collocation software for boundary-value codes, ACM, Trans. Math. Software, 7, 209–222, 1981.

Ascher, U. M., R. M. M. Matheij, and R. R. Russell, Numerical solution of boundary value problems of ordinary differential equations, Prentice Hall, Englewood Cliffs, NJ, 1988.

Azia, K. and Settari, A., Petroleum reservoir simulation, London: Applied Science, 1979.

Babuska, I., O. C. Zcienkiewicz, J. Gago and E. R. de A. Oliveira, Accuracy estimates and adaptive refinements in finite element computations, John Wiley & Sons, 1986.

Babuska, I. and W. Rheinboldt, Error estimates for adaptive finite element computations, SIAM J. Numer. Anal. 15, 736–545, 1978.

Bader, G. and U. M. Ascher, A new basis implementation for a mixed order boundary value code solver, SIAM J. Sci. Statist. Comput., 3, 483–500, 1982.

Baines, M. J. and A. J. Wathen, Moving finite element methods in evolutionary problems. I. Theory, J. Comp. Phys., 79, 255–269, 1988.

Baker, A. J. and J. W. Kim, A Taylor weak-statement algorithm for hyperbolic conservation laws, Int. J. Num. Methods Fluids, 7, 489–520, 1987.

Bear, J., Dynamics of fluids on porous media, Elsevier, New York, 1972.

Bear, J., Hydraulics of groundwater, McGraw-Hill Inc., New York, 1979.

Beck, J. V. and Arnold, K.V., Parameter estimation in engineering and science, John Wiley & Sons, New York, 1977.

Benner, E. R., H. T. Davies, and L. E. Scriven, An adaptive finite element method for steady and transient problems, SIAM J. Sci. Statist. Comp., 8, 529–547, 1987.

Berger, M. and Oliger, J., Adaptive mesh refinement for hyperbolic partial differential equations, J. Comput. Phys. 53, 484–512, 1984.

Buckley, S. E. and Leverret, M. C., Mechanisms of fluid displacement in sand, Trans. AIME, 146, 107–116, 1942.

Carrea, J., and S. P. Neuman, Estimation of aquifer parameters under transient and steady state conditions, 1., 2. and 3., Water Resour. Res., 22, 199–242, 1986.

Coats, K., An equation-of-state compositional model, Soc. Petr. Eng. J., 363–376, 1980.

Conder, J.R., and C. L. Young, Physicochemical measurement by gas chromatography, John Wiley & Sons, 1977.

Concus, P. and W. Proskurowski, Numerical solution of a nonlinear hyperbolic equation by the random choice method, J. Comput. Phys., 30, 153–166, 1979.

Donea, J., S. Giuliani, H. Laval and L. Quartapelle, Time-accurate solution of advection diffusion problems by finite elements, Comput. Meths. Appl. Mech. Eng., 45, 123–145, 1984.

Donea, J., L. Quartapelle and V. Selim, An analysis of time discretization in the finite element solution of hyperbolic problems, J. Comp. Phys., 70, 463–499, 1987.

Ewing, R. E. (ed.), The mathematics of reservoir simulation, Society for Industrial and Applied Mathematics, Philadelphia, 1983.

Ewing, R. E., Efficient adaptive procedures for fluid-flow applications, Comput. Meths. Appl. Mech. Eng., 55, 89–103, 1986.

Ewing, R. E., Characteristic Petrov-Galerkin methods of advection-dominated flows, in: Proceedings of the Seventh International Conference on Finite Element Methods in Flow Problems, Huntsville, Al, April 3–7, 1989, T.J. Chung and G. R. Kerr (eds.), UAH Press, Huntsville, Al, 1989.

Fletcher, R., Practical methods of unconstrained optimization, John & Wiley, New York, 1980.

Fletcher, C.A.J., Computational techniques for fluid dynamics, Vol.I, Springer, New York, 1988.

Gray, W.G. and G. F. Pinder, An analysis of the numerical solution of the transport equation, Water Resour. Res., 12, 547–555, 1976.

Gebhart, B., Y. Jaluria, R. L. Mahajan, and B. Sammakia, Buoyancy-induced flows and transport, Hemisphere Publishing Corporation, New York, 1988.

Gill, P. E., Murray, W. and M. H. Wright, Practical optimization, Academic Press, Orlando, San Diego.

Gupta, S.P. and R.A. Greenkorn, Dispersion during flow in porous media with bilinear adsorption, Water Resour. Res., 5, 1357–1368, 1973.

Gropp, W. A., Local uniform mesh refinement with moving grids, SIAM J. Stat. Comp., 8, 292–304, 1987.

Harten, H., High resolution schemes for hyperbolic conservation laws, J. Comput. Phys., 49, 357–393, 1983.

Hughes, T. J. R. (ed.), Finite element methods in convection-dominated flow problems, ASME, AMD-34, New York, 1979.

Javandel, I., C. Doughty and C. F. Tsang, Groundwater transport: handbook of mathematical models, Water resources series monograph series 10, American Geophysical Union, Washington, DC, 1985.

Johnson, I. W., A. J. Wathen and M. J. Baines, Moving finite element methods in evolutionary problems, II, Applications, J. Comp. Phys., 79, 270–297, 1988.

Karickhoff, S. W., Semi-empirical estimation of sorption of hydrophobic pollutants on natural sediments, Chemosphere, 10, (8), 833–846, 1981.

Kee R. J., and L. A. Petzold, A differential/algebraic equation formulation of the method of lines solution to system of partial differential equations, in Advances in computer methods in partial differential equations, VI, Vichnevsky, R. and S. Stepleman, eds., IMACS, 1987.

Koch, M. Optimal regularization of the linear seismic inverse problem, Proceedings of the SIAM workshop on 'Geophysical inversion', Houston, Tx, Sept, 20–25, 1989, Society for Industrial and Applied Mathematics, Philadelphia, 1991.

Koch, M. Dissipation/dispersion minimizing numerical solution of hyperbolic equations by an adaptive method of line, to be submitted to J. Comp. Phys., 1990a.

Koch, M., Numerical investigation of two-phase Buckley-Leverret flow in the unsaturated zone by an adaptive method of line, to be submitted to Water Resour. Res., 1990b.

Koch, M. and G. Zhang, Development of an analytical model for the determination of solute transport in groundwater, Report to the Florida Department of Environmental Regulation, Tallahassee, Fl, 1989.

Koch, M. and G. Zhang, A comparison of some numerical techniques for the solution of advection dominated transport problems, to be submitted to Int. J. Num. Meth. Fluids, 1990.

Knopmann, D. S. and C. I. Voss, Behavior of sensitivities in the one-dimensional advection-diffusion equation: Implications for parameter estimation and sampling design, Water Resour. Res., 23, 253–272, 1987.

Lapidus, L. and Amundson, N. R., Mathematics of adsorption in beds, VI, The effect of longitudinal diffusion in ion exchange and chromatographic columns, J. Phys. Chem., 56, 984–988, 1952.

Lapidus, L., and G. Pinder, Numerical solution of partial differential equations in science and engineering, John Wiley & Sons, 1982.

Liskovets, O. A., The method of lines (review), Differential Equations, 1, 1308–1323 (translated from Russian), 1965.

Levenberg, K., A method for the solution of certain nonlinear problems in least squares, Quart., Appl. Math., 2, 164–168, 1944.

Loehner, R, K. Morgan, and O. C. Zienkiewicz, The solution of nonlinear systems of hyperbolic equations by the finite element method, Int. J. Num. Methods Fluids, 4, 1043–1063, 1984.

Luckner, L., M.Th. van Genuchten, and D.R. Nielsen, A consistent set of parametric model for the two-phase flow of immiscible fluids in the subsurface, Water Resour. Res. 25, 2187–2194, 1988.

Marquardt, D.W., An algorithm of least squares estimation of nonlinear parameters, J. Soc. Indust. Appl. Math., 11, 431–441, 1963.

Miller, K., Moving finite elements, II, SIAM J. Numer. Anal, 18, 1033–1057, 1981.

Miller, K. and R. Miller, Moving finite elements, I, SIAM J. Numer. Anal, 18, 1019–1032, 1981.

Navon, I.M. and D.M. Legler, Conjugate-gradient methods for large-scale minimization in meteorology, Monthly Weather Review, 115, 1479–1502, 1987.

Oden, J. T., Demkowicz, L., Strouboulis, T., and P. Devloo, Adaptive methods for problems in solid and fluid mechanics, in: Accuracy estimates and adaptive refinements in finite element computations, John Wiley & Sons, 1986.

Parker, J. C., Multiphase flow and transport in porous media, Rev. Geophys., 27, 311–328, 1989.

Petzold, L., A., Adaptive moving grid strategies for one-dimensional systems of partial differential equations, in Advances in computer methods in partial differential equations, VI, Vichnevsky, R. and S. Stepleman, eds., IMACS, 1987.

Satter, A., Y.M. Shum, W.T. Adams, and L.A. Adams, Chemical transport in porous media with dispersion and rate-controlled adsorption, Soc. Pet. Eng. J., 129–13, Feb. 1980.

Sewell, G., The numerical solution of ordinary and partial differential equations, Academic Press, San Diego, 1988.

Shanno, D.F. and K.H. Phua, Remark on algorithm 500–a variable method subroutine for unconstrained optimization, ACM Trans. on Mathematical Software, 6, 618–622, 1980.

Sod, G. A., Numerical methods in fluid dynamics, Initial and initial boundary-values problems, Cambridge University Press, Cambridge, 1985.

Tezduyar, T.E. and D. K. Ganjo, Petrov-Galerkin formulations with weighting functions dependent upon spatial and temporal discretization: applications to transient convection-diffusion problems, Comput. Methods App. Mech. Eng., 59, 49–71, 1986.

Van Genuchten, M. Th., A closed-form equation for predicting the hydraulic conductivity of unsaturated soils, Soil Sci. Soc. Am. J., 44, 892–898, 1980.

Van Genuchten, M. Th. and W. J. Alves, Analytical solutions of the one-dimensional convective-dispersive solute transport equation : U.S. Department of Agriculture Tech. Bull., 1661, 1982.

Wagner, B. J. and S. M. Gorelick, A statistical methodology for estimating transport parameters,: Theory and applications to one-dimensional advective-dispersive systems, Water Resour. Res.,22, 1303–1315, 1986.

Warming R. F. and B. J. Hyett, The modified equation approach to the stability and accuracy analysis of finite-difference methods, J. Comp. Phys., 14, 159–179, 1974.

Wheeler, M., (editor) Numerical simulation in oil recovery, Springer Verlag, Berlin, New York, 1988.

Yeh, W. W-G., Review of parameter identification procedures in groundwater hydrology: The inverse problem, Water Resour. Res., 22, 95–108, 1986.

CHAPTER 12

Calculating Flow and Transport in Porous/Fractured
Media Using the Cellular Automata Approach

Bryan J. Travis*
Kenneth G. Eggert*
Shi Yi Chen**
Gary Doolen**

INTRODUCTION

A unique feature of flow and transport in soils and rocks is the highly multiply-connected topology and varied channel geometries through which fluids and solutes must pass. Phenomenological and averaged models have had considerable success in describing porous flow and transport, but a truly predictive capability will not exist until microphysical (pore scale) details can be modeled and well characterized and coupled to larger scale dynamics.

To model flow and transport on a microphysics scale, two things are necessary: (1) an efficient means of generating complex 2-D and 3-D multiply-connected structures; and (2) a highly efficient method for solving flow and transport equations in arbitrary geometries. The ability to perform these two tasks may now exist. The concept of fractal dimension has led to the development of fast algorithms for generating rough surfaces and realistic, sponge-like solids, as theoretical models of porous media. Also, photomicrographs of rock thin sections and x-ray tomography scans of rocks provide access to the detailed structure of porous materials in 2 and 3 dimensions. These images can be digitized and the patterns can then be etched into glass, providing experimental analogues of real materials. The same digitization can be used to describe pore geometries for numerical simulations. Flow and transport in these very complex structures can be calculated with the emerging discipline of cellular

*Earth and Environmental Sciences Division, Los Alamos National Laboratory, Los Alamos, NM 87545.
**Center for Nonlinear Studies, Los Alamos National Laboratory, Los Alamos, NM 87545.

automata, providing not only permeability values, but also the possibility of generating highly accurate constitutive relationships for use in large scale field applications. This paper focuses on the use of cellular automata models of fluid flow in porous media of arbitrary complexity.

This paper first describes the cellular automata approach to simulating fluid dynamics. Then several applications of the method are presented for calculation of (1) permeability as a function of porosity and fractal dimension, (2) simulation of flow in etched glass micromodels, (3) dispersion as a function of fractal dimension; and (4) fracture flow.

Cellular Automata

Traditionally, mathematical treatments of fluid flow have taken the continuum approach. Flow is considered as a transport of some smooth, infinitely fine quantity (mass, momentum, or energy). Dynamic equations are derived by considering the net flux through a small volume---there is little or no reference to the underlying atomistic nature of the flow. This approach has been extremely successful in describing a wide variety of fluid flow situations. However, the resulting continuum equations are highly nonlinear. Numerical solutions or perturbation expansions must often be sought because of the difficulty or impossibility of finding analytic solutions.

Numerical solutions involve a discretization of the continuum into a discrete micromodel of the flow in which parcels of fluid move between small, well-defined volumes. It has occurred to many researchers that a ``bottom-up'' approach, using a discrete, atomistic model of fluids to start with might be a more efficient path than the continuum. The difficulty with this idea is that a direct model of molecular dynamics would require far too many calculations to be feasible. A discrete method in which particles represent collective behavior of many atoms or molecules is needed.

Discrete fluid models have undergone an evolution in which the level of detail and the essential features needed to reproduce Navier-

Stokes behavior on the macroscale have crystallized. Cellular automata, collections of particles which move on a lattice according to a set of propagation and interaction rules, are capable of complex behavior and can reproduce fluid flow dynamics under the right conditions. Frisch, Hasslacher, and Pomeau (1986) have recently determined what those conditions are. Their model is hereafter referred to by the initials FHP.

Minimal requirements for obtaining Navier-Stokes behavior are few and simple:

(1) A collection of N identical particles moving with uniform speed

$$M_i = M_0, \; |\vec{V_i}| = C_0, \; 1 \le i \le N \tag{1}$$

A discrete phase space, that is, the locations of particles are quantized

$$\vec{x_i} = (x_k, y_j) = \{ x_k = x_0 + k\Delta x, \; y_j = y_0 + j\Delta y \} \tag{2}$$

and time is also uniformly discretized

$$t_s = s \, \Delta t \tag{3}$$

(3) All particles are constrained to move on a hexagonal lattice in 2-D, i.e., the velocity directions are also quantized---

$$\tan^{-1}\left(\frac{v_y}{v_x}\right) \varepsilon \{ 0^\circ, 60^\circ, 120^\circ, 180^\circ, 240^\circ, 300^\circ \} \tag{4}$$

(4) Particles can interact through binary and symmetric triple collisions, in which mass and momentum must be conserved. This can be achieved by rotating particles ±60 through collision points, for example (see Fig. 1).

5) Particles obey an exclusion principle, i.e.,

$$\vec{V}_{p_1}(i,j) \neq \vec{V}_{p_2}(i,j) \tag{5}$$

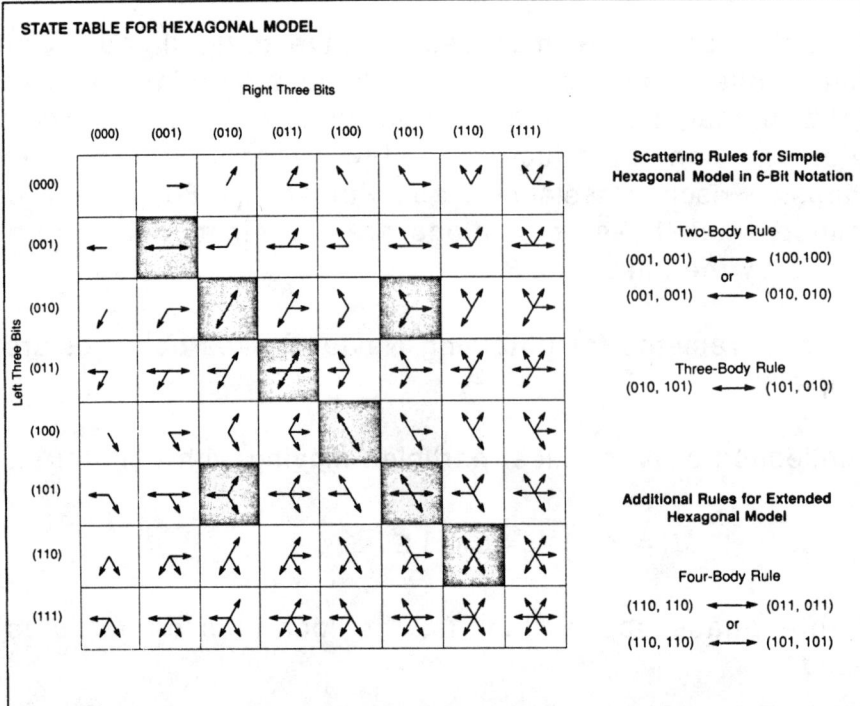

Figure 1a. All possible states of the hexagonal lattice. Each state can be expressed in 6-bit rotation. Gray shaded states correspond to extended hexagonal simple model. (From Hasslacher, 1987).

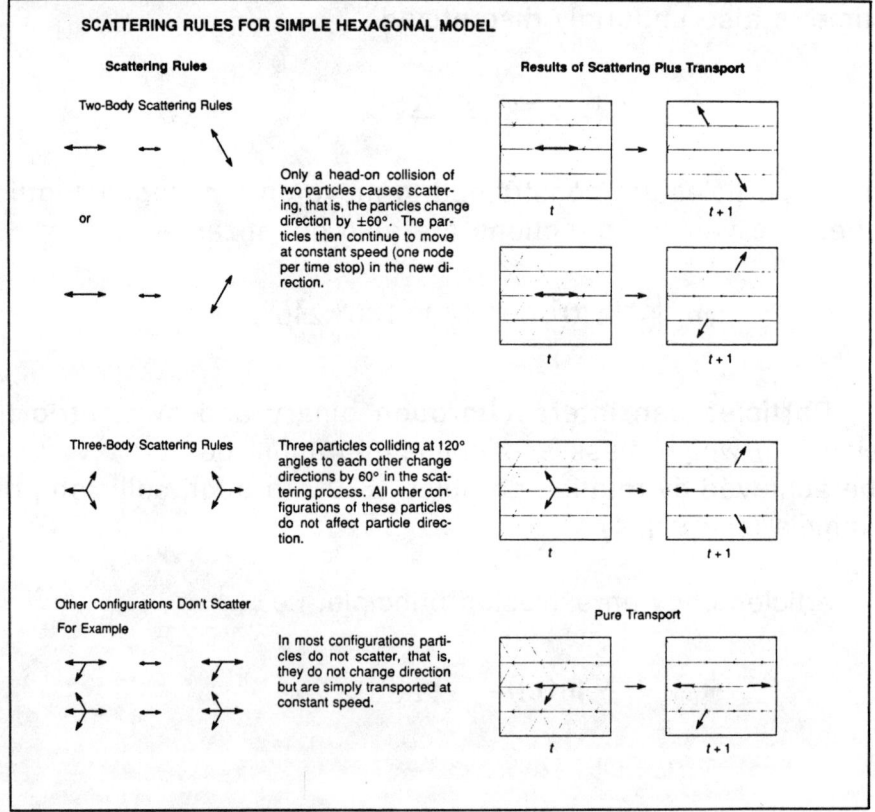

Figure 1b. Scattering rules for simplest model.

No two particles (P_1, P_2) at a node can have the same speed and direction. This requires that the number of particles at a node cannot exceed 6.

More sophisticated models can be constructed, e.g., by expanding the collision set to include asymmetric 3-body interactions and higher order collisions or by adding particles that move at a different speed. These more complex models provide greater accuracy for a given lattice discretization and alloe higher Reynolds number flows to be simulated, but require more computer time.

In the limit of an incompressible, low Mach number ($M = V/c_s$ << 1) lattice gas, the macroscale dynamics is given by the lattice Navier-Stokes equation

$$\rho_0 \, \partial_t V + \rho_0 g(\rho_0) \, V \cdot \nabla V = -c_s^2 \, \nabla \rho + \rho_0 \nu \, (\rho_0) \, \nabla^2 V \tag{6}$$

and

$$\nabla \cdot V \;=\; 0 \tag{7}$$

To recover Galilean invariance, time and viscosity and pressure must be scaled:

$$t \to \frac{t'}{g(\rho_0)}, \quad \nu \to g(\rho_0)\nu' \tag{8a}$$

and

$$p \to \frac{\rho_0 g\,(\rho_0)}{c_s^2}\, p' \tag{8b}$$

Equation (7) then is transformed to the standard incompressible Navier-Stokes equation

$$\partial_t \cdot V' + V' \cdot \nabla\, V' = -\nabla P' + \nu' \, \nabla^2 V' \tag{9}$$

The model we use is a six-bit version of the FHP scheme. Our model includes a 7th particle or bit type to identify nodes that are in the solid matrix of a porous formation. (For an excellent summary the reader is referred to the article by Hasslacher (1987). For details on derivations of Navier-Stokes equations from CA models and for other technical information, the Volume 1, No.4, 1987 issue of *Complex Systems Journal* is a good starting point.)

One important consideration in CA calculations in porous media is the mean free path length relative to pore/channel width. If too few nodes are used in a region, the CA method cannot give an accurate approximation to the Navier-Stokes equations.

The mean free path in CA depends on both the particle density at a node as well as the family of collisions that have been included. The mean free path $\langle l \rangle$ is given by

$$\langle l \rangle = \frac{c_s \tau}{f} = \frac{1}{f} \tag{10}$$

where $c_s \tau = 1$ is the internode separation, and f is the frequency of collisions at a node, and is given by

$$f = \sum_n n f_n \tag{11}$$

Here f_n is the probability of an n-body collision, given by

$$f_n = k_n d^n (1-d)^{6-n} \text{ for } 2 \leq n \leq 5 \tag{12}$$

and k_n is the number of possible configurations in which an n-body collision can occur (e.g., in a 3-body collision, the particles can collide symmetrically; i.e., 120° between paths, or with a head-on collision between two with the third entering from one of the other four paths, etc.) and d is the average particle number density at a node.

For example, in the minimal CA model (2-body collisions and symmetric 3-body collisions), and for density $d = 0.2$, $f \cong 0.12286$ and $\langle l \rangle \cong 8.14$ nodes. When nonsymmetric 3-body collisions are included, f increases to 0.27 and $\langle l \rangle \cong 3.70$. Addition of some 4-body collisions further reduces $\langle l \rangle$ to 3.54. The CA model used here is the 6 bit, collision-saturated deterministic version (Diemer et al, 1989), designated here as CA6CSD, which includes all 2-, 3- and 4-body collisions that can result in a output state different from the collision input configuration.

The mean free path can also be reduced by using a larger density d.   For example, using CA6CSD, $d = 0.33$ yields $\langle l \rangle = 1.54$ while $d = 0.47$ gives $\langle l \rangle = 1.09$. At high densities (near 0.5), the proportions of collisions that are 3-body or higher interactions increases (Fig. 2) and to maintain accuracy only models including all possible collisions. should be used.   Also, the fluid viscosity will change considerably with density.   Fluid viscosity for the minimal six-bit FHP model is given by Frisch et $al.$, (1986)

$$V = \frac{1}{12} \frac{1}{d\,(1-d)^3} - \frac{1}{8}$$

(13)

Other important quantities are the node density

$$\rho = 6d,$$

(14)

the scaling factor,

$$g(\rho) = \frac{3-\rho}{6-\rho} = \frac{1}{2}\left(\frac{1-2d}{1-d}\right)$$

(15)

mass density

$$\rho' = \frac{2\,g\,(d)}{\sqrt{3}\,l^2}$$

(16)

and the sound speed

$$C_s = \frac{1}{\sqrt{2}}$$

(17)

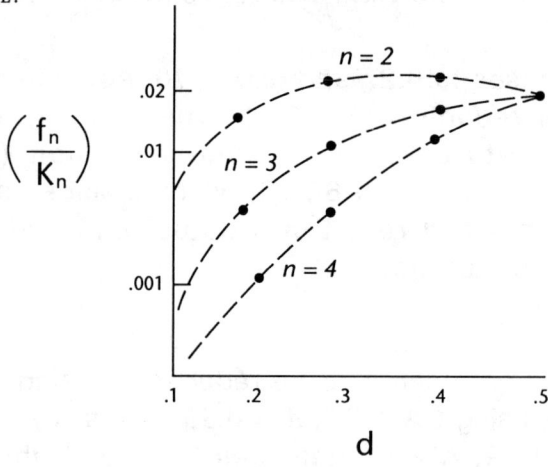

Figure 2. Proportion of n-body collisions expected as a function of particle density d for different n.

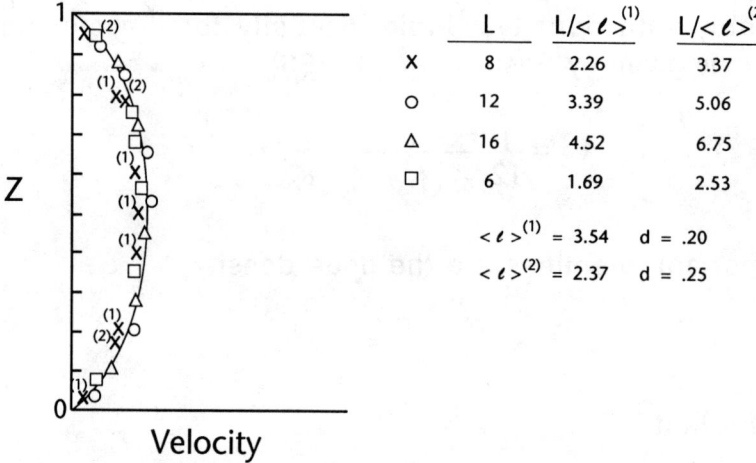

	L	$L/<\ell>^{(1)}$	$L/<\ell>^{(2)}$
X	8	2.26	3.37
O	12	3.39	5.06
△	16	4.52	6.75
□	6	1.69	2.53

$$<\ell>^{(1)} = 3.54 \quad d = .20$$
$$<\ell>^{(2)} = 2.37 \quad d = .25$$

Figure 3a. Comparison of calculated velocity profile for various discretization vs. analytic solution for viscuous flow in a uniform channel.

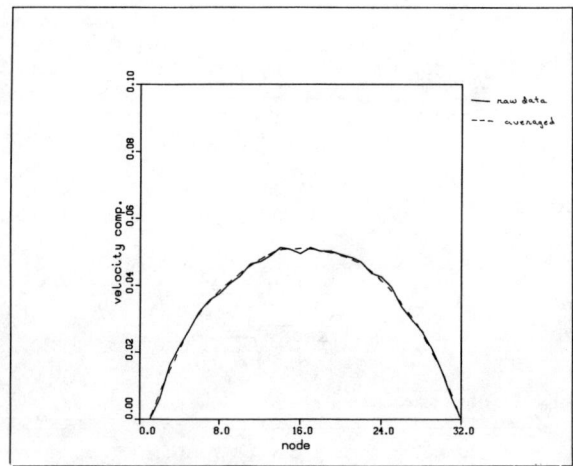

Figure 3b.

For our 6-bit collision-saturated model, eqns. (13) and (15) change (Diemer et al, 1989) to:

$$v = \frac{1}{8} \left[\frac{1}{3\,d\,(1-d)^3 + 12\,d^2(1-d)^2 + 3\,d^3\,(1-d)} - \frac{1}{2} \right] \tag{18}$$

and

$$g = \frac{1}{4}\,\sqrt{2}\,\left(\frac{1-2d}{1-d}\right). \tag{19}$$

As we have seen, by including most, if not all, of the collision types, and by using a moderate to high density, the mean free path can be made quite small---almost one node spacing long. In the Boltzmann approximation, it was assumed that $\langle l \rangle \ll L$, where L is a characteristic length of the system. In fact, $\langle l \rangle$ can be on the order of L and the flow system will still give a reasonable approximation to the Navier-Stokes equations. Figure 3 compares the velocity profiles obtained for different $L/\langle l \rangle$ ratios. We can compute flow in channels which are only 5 or 6 nodes wide.

Another matter of importance for the CA approach to porous flow is the problem of noise. Averaged (microscopic) quantities such as velocity and pressure experience fluctuations due to the atomistic nature of the simulation causing fluctuations in particle directions at nodes. The magnitude of these fluctuations can be controlled by changing the space and time intervals over which the averaging process is performed. The noise level $\delta V / V$ in velocity varies as (Dahlburg, Montgomery and Doolen, 1987)

$$\frac{\delta V}{V} \approx \frac{1}{\sqrt{NT}} \tag{20}$$

where N is the number of nodes in the spatial averaging region and T is the time interval (number of time steps). In steady flow calculations, we generally take $N = 1$ and T is a large number, on the order of $10^{**}4$. This results in a 1% noise level. For transient flow, averaging over some number of nodes is required to maintain an acceptably low noise level. Fig. (3.b) illustrates the velocity profile in a long channel with no nodal averaging. The fluctuations in velocity are small but noticeable. However, the profile follows the expected form on average. Fig. (3.b) also shows the velocity profile for the same situation with averaging over the neighboring 6 nodes.

When the Navier-Stokes equations are cast in dimensionless form, the Reynolds number (Re) is the characterizing parameter. We can use Re to convert our CA solutions to dimensional quantities.

$$Re = V L \frac{g(\rho)}{v(\rho)} = M L \left(\frac{c_s g(\rho)}{v(\rho)} \right) = M L R^* \tag{21}$$

where R^* is a lattice Reynolds number, M is the Mach number and L is a global length scale.

For $d = 0.2$, using CA6CSD, we get

$$Re \cong 0.492 \, M L \tag{22}$$

Applications

We describe in this section several applications of the cellular automaton method to flow and transport in porous media. These examples address several important problems, namely, the calculation of permeability as a function of pore structure, flow in fractures, and the calculation of contaminant transport.

Flow in porous media is generally modelled using Darcy's law, which states that mass flux is directly proportional to pressure gradient and matrix permeability (cf, Bear, 1972). This is an averaged equation, i.e., it relates bulk flow to macroscopic pressure. The permeability factor depends on the details of pore structure and connectivity in some complex way. Many attempts have been made to predict permeability as a function of pore size distributions, pore shapes, porosity, and saturation (in multi-phase flows) using various simplifications, ranging from simple bundles of capillary tubes to complex networks of channels and pores (cf. Dullien, 1979). These permeability models are generally of limited application, being valid only for certain kinds of rock, and over limited porosity ranges.

Rothman (1988) has previously shown that CA techniques can be used to simulate flow in porous media. However, the models of porous materials he used were idealized as channels passing around blocks, in a somewhat regular fashion. We show here that more complex and realistic geometries can also be treated.

Generation of multiply connected porous structures

First, we demonstrate a relationship between permeability and porosity and fractal dimension. A series of calculations have been performed, each computing flow through a fractal porous medium. The creation of 2-D and 3-D structures having complex but fractal (i.e., statistically self-similar) internal connectivity is achieved in a simple manner. The basic procedure is to distribute features (either grains or open spaces) randomly in a region, using an appropriate number density distribution function. Grain and pore sizes for rocks and soils frequently follow a log-normal distribution; however, it has been observed (e.g., Krohn, 1988) that many rocks have a fractal distribution of features at small scales (less than a hundred microns or so). In a fractal medium, the number density n of features of size L is given by

$$n(L) = A \, L^{**}(2-D) \tag{23}$$

where D is the fractal dimension and A is a coefficient which varies with the material. In this section, we restrict our simulated media to those with fractal characteristics only. Our model structure generator will, however, allow a log-normal distribution for larger features as well as the fractal law for small to intermediate scales. The following procedure is used.

Step 1. Pick a set of feature sizes. The sizes must lie in in the range of validity of (23). Typically, each successive size is picked by diminishing the previous size by a factor of two.
Step 2. Determine the number density for each feature size from equation (23). Calculate $\Delta = \Delta x = \Delta y = \Delta z = 1/n(L)$. We now divide the region of interest into subdomains of size $N^{*}\Delta x$, $N^{*}\Delta y$, $N^{*}\Delta z$, where the value of N is supplied by the user. (The overall dimensions of the region must be whole multiples of $N^{*}\Delta$). The expected number E(L) of features of size L in each subdomain is $N^{**}3$.

Step 4. Within the specified subdomains, the expected number of features of size L are distributed randomly. Feature shapes can be cubes or spheres. The largest feature is distributed first. Smaller features can overlap the already distributed larger ones. We restrict the locations of features to the subdomains rather than to the entire region. This will tend to restrict the possible realizations, especially if smaller subdomains are chosen, mainly by eliminating those with extreme variations due to large-scale clumping of features. The structures we get should be typical in a sense. The values of Δx, Δy and Δz are usually considerably larger than the corresponding L; so the restriction to placement within subdomains is not a particularly strong one.

Step 5. The nature of any feature, i.e., whether it represents a solid grain or an open region, is determined probabilistically. Initially, all nodes are set to grain. An input quantity e, ranging between 0 and 1, is used to control the final porosity. A feature is declared a solid grain if

$$\min(1, \text{int}(\text{rand}/e)) = 1 \tag{24}$$

otherwise, it represents open space. In eqn. (24), 'rand' is a random number between 0 and 1, and 'int' is the greatest integer function. With this procedure, porosity of the final structure is invariably quite close to e. In 3-D, structures containing at least one path through the system are obtained for realistic values of porosity, down to about 10-15 % (Chen et al., 1990). In 2-D, we frequently cannot generate connected systems for porosities below about 45 - 50 %.

A square 2-D region is discretized by a regular hexagonal lattice of 256 nodes in each direction (x and z). The upper and lower surfaces are no-flow boundaries. Fixed pressures are specified on the left and right sides, with the left side pressure larger than on the right side. Pressures are generated by injecting more particles (larger d) on the left than on the right. A value of d= 0.20 is assigned to the right boundary and initially at all interior nodes. A value of d = 0.22 is assigned to the left boundary nodes. Simulations are run for 30-50,000 cycles or more to insure that steady state has been reached.

Figures 7, 8, and 9 illustrate the steady state velocity fields corresponding to the fractal media displayed in Figs. 4, 5, and 6, respectively.   At the lower values of D , flow paths are less tortuous and pore throats are larger; flow moves through several channels (Fig. 7).   At intermediate values of $D \approx 2.50$, average flow rates are lower, flow paths are more tortuous and there are fewer prominent channels (Fig. 8).   At high values of D (> 2.7), pores are smaller, flow resistance is considerably higher (Fig. 6).   The flow structure also appears more isotropic; we do not see a few primary flow paths, but many weak ones (Fig. 9).   In all three cases, regions of essentially stagnant flow are seen as well as dead-end pores and channels.

Figure 10 displays the pressure field corresponding to Figs. 4 and 7. Black regions represent the grains of the soil. Dark shades of gray correspond to higher pressures, light shades to low pressures. Most of the pressure drop occurs at the pore throats.

The permeability of these structures can be computed from

$$k = \frac{Q \mu \Delta L}{A \Delta P} = \frac{M_{in} V \rho L_x}{L_z \cdot \Delta d}$$

(25)

where Q is mass flux, Min is the number of particles entering a model, V is average horizontal velocity, ρ is fluid density, Lx and Lz are dimensions in the horizontal and vertical coordinates, respectively, μ is fluid viscosity, A is cross-sectional area, and ΔP and. Δd are pressure and particle density differences across the sample.

Results of several dozen calculations are plotted in Fig. 11.   As expected, the less tortuous structure of the low D media results in higher permeabilities than for the higher D cases. Variations in permeability for different realizations at constant D and ε are generally less than 5%.   Additional cases have been run using different pore size distributions.   These cases, labelled PSD2 and PSD3, contain finer particles.   As seen in Fig. 11, permeability decreases as the size of the smallest particles decreases, in keeping with experimental observations.   Also shown in Fig. 11 is the increase in permeability as a function of porosity for fixed $D = 2.5$.

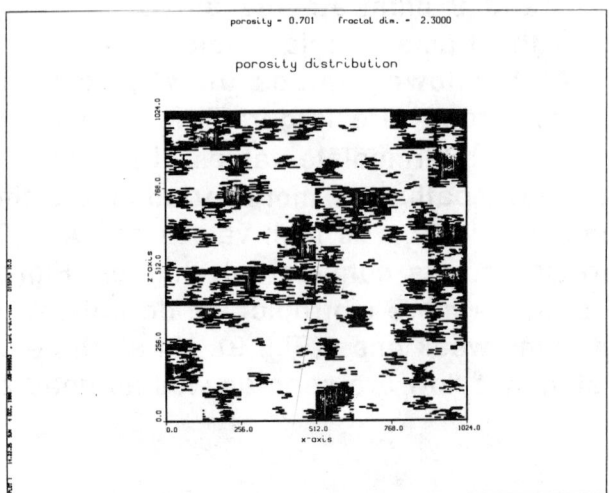

Figure 4.

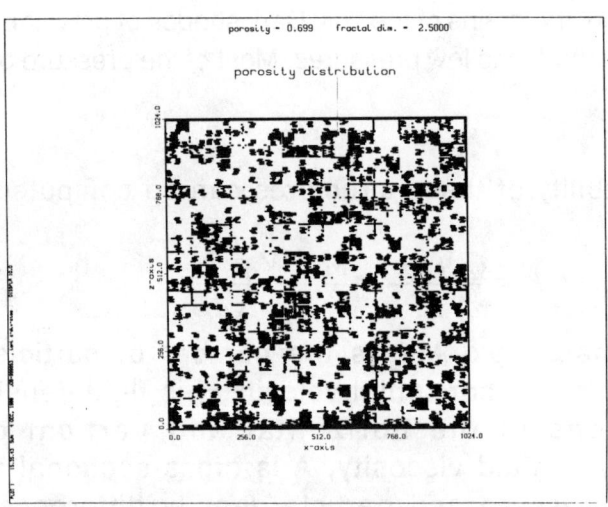

Figure 5.

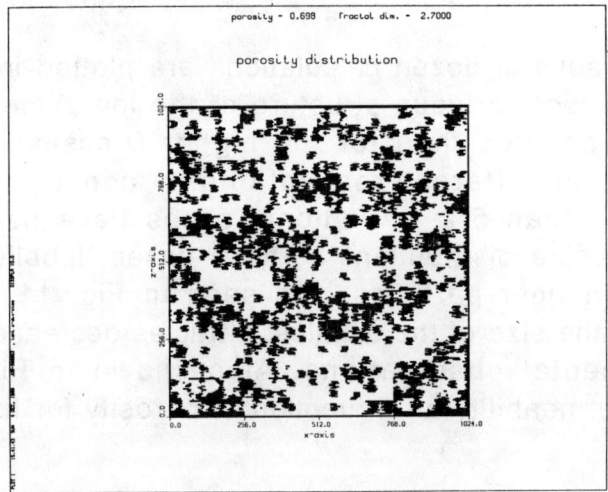

Figure 6.

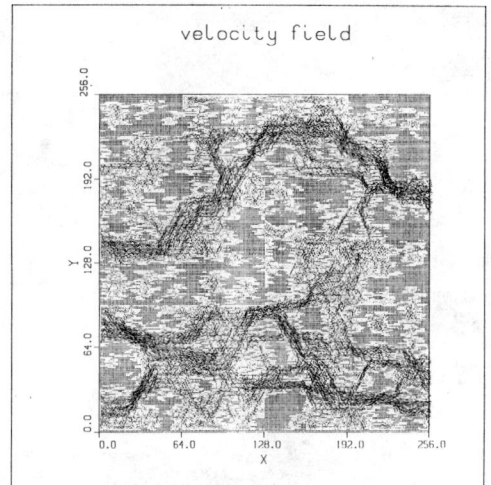

Figure 7.

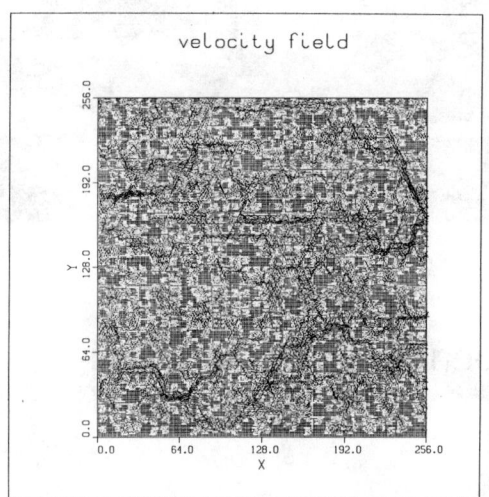

Figure 8.

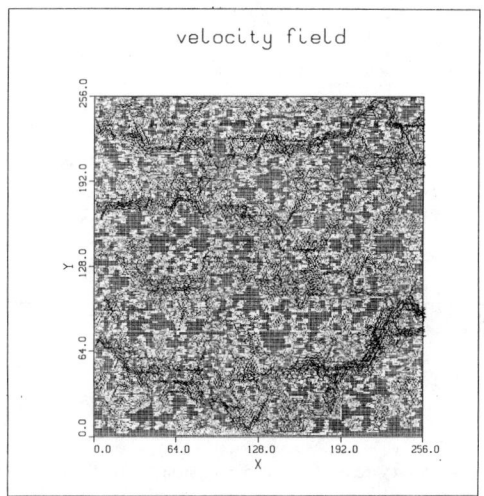

Figure 9.

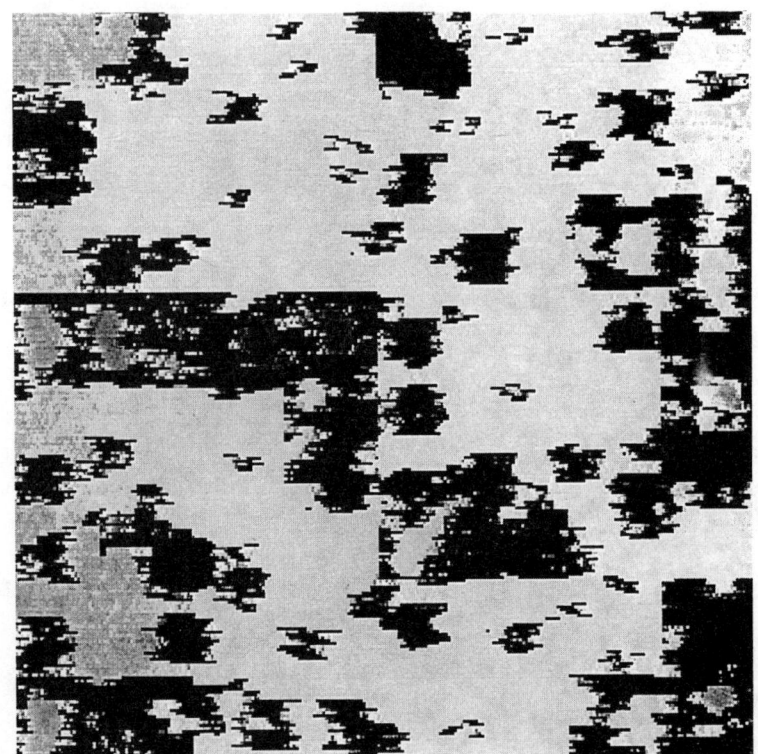

Figure 10.

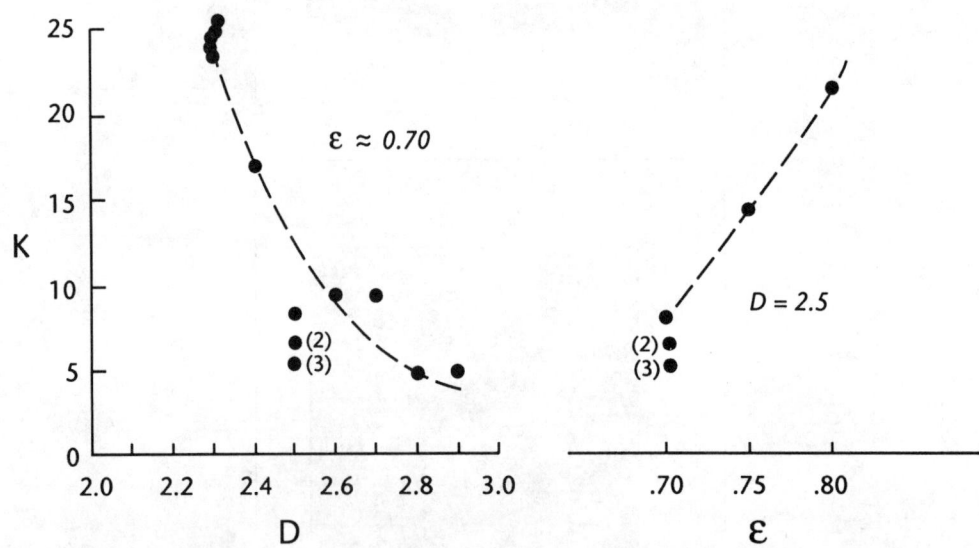

Figure 11. Dependence of conductivity K on fractal dimension D for fixed porosity ϵ (left) and on a limited range of porosity for a fixed fractal dimension.

A truly general permeability method would be available if the pore scale structure could be determined routinely for any porous material and if fluid flow through that geometry could be calculated easily. Until recently, these two criteria seemed out of reach. Now, digitization of microphotographs of thin sections of rocks and even x-ray tomography provide access to the detailed structure of porous materials in 2 and 3 dimensions.

Photographs of thin sections of carbonate rocks can be digitized (e.g., Bahralolom, Bretz and Orr, 1988). Using a method similar to printed circuit fabrication, these digitized images are then etched into glass, providing an accurate model of the porous sample (at least in 2 dimensions). Flow and transport experiments can be performed on these etched glass systems. Because they are made of glass, the interior flow is fully visible. These models have seen considerable use for multiphase flow studies as well as for single phase flow and transport.

The digitized images used for these etched glass models can be used directly in our calculations. Two examples are given here which illustrate the utility of cellular automata (CA) methods in complex porous media. Figure 12a shows an example of an image for an etched glass experiment. The dark objects are the matrix grains. Grain sizes follow a narrow size distribution. Physical dimensions of this etched glass model are 6.4 cm long, and 4.3 cm wide. The etchings are 100 microns deep on average. The porous structure is surrounded by an impermeable region. Fluid is injected through a narrow port on the left and withdrawn at the right. In our CA simulations, pressures are generated by injecting more particles (larger d) on the left than on the right. A value of $d = 0.25$ is assigned to the right boundary and initially at all interior nodes. A value of $d = 0.35$ is assigned to the left inlet nodes. Simulations are run for 30-50,000 cycles or more to insure that steady state has been reached. A lattice of 1200 by 2048 nodes were used.

The permeability of these structures can be computed from eqn. (25). The measured permeability of this fairly homogeneous system is approximately 23 darcys. Our numerical simulations yielded a permeability of 21.9 darcys.

A second etched glass sandwich has also been simulated. (Fig. 12b). This model is much more heterogeneous than the first one considered. It has a fairly well connected channel passing through it, and several large void spaces. Physical dimensions of this etched glass model are 7.0 cm long, and 5.0 cm wide. The etchings are 50 microns deep on average. It has a measured permeability of 16.1 darcys. A lattice of 1144x2048 nodes were used in the CA simulation of this model. Calculated permeability is 15.6 darcys.

Transport

Prediction of solute transport and dispersion in porous media is of great interest. On a microscopic level, velocity fields can be quite tortuous. This strong fluctuation on a small scale tends to spread a concentrated parcel of contaminant over a large region, in the transverse direction as well along the principal direction of flow. Our CA model has been adapted to follow particles of a different "color" through the flow fields. By keeping track of where these particles exit and when, we can build up a set of data for determining dispersion coefficients in porous media. Our initial goal will attempt to relate dispersivity to fractal dimension, as was done for permeability. Figs. 13 and 14 are snapshots from two simulations of tracer migration using the geometry of Fig. 4 and the velocity field of Fig. 7. The location of the injection point on the left boundary is the only difference. As the particles are carried along with the transporting fluid, they experience small scale deflections and disperse. Most of the particles flow along the main channels to the exit; however, some have migrated into other regions of the porous formation, and still others are virtually trapped in dead-end pores or very low speed, narrow channels.

Fig. 15 plots the results of several simulations of tracer movement vs. fractal dimension D. Transverse disersivity has been normalized by the value in a uniform channel with the same average flow rate as the fractal medium has. Two calculations, with different injection points, are performed for each geometry and results are averaged. As fractal dimension increases, the dispersivity coefficient increases. The more tortuous geometry at higher D values is more efficient at spreading solute.

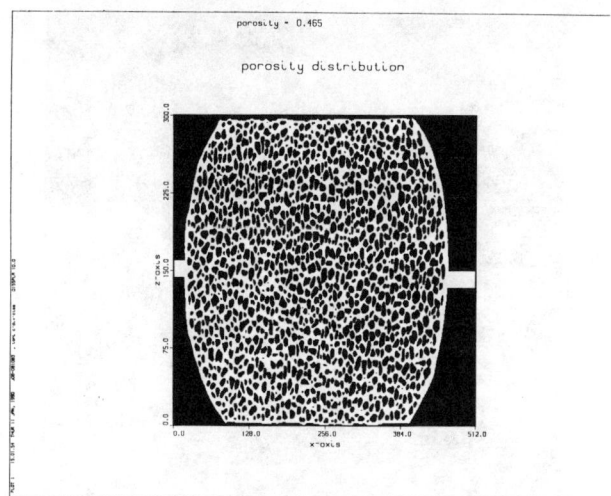

Figure 12a.

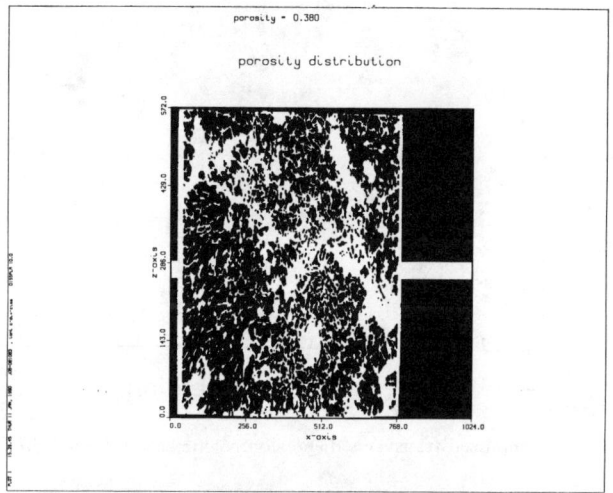

Figure 12b.

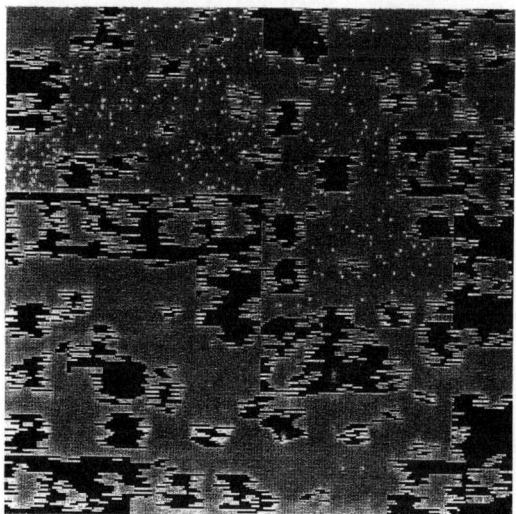

Figure 13.

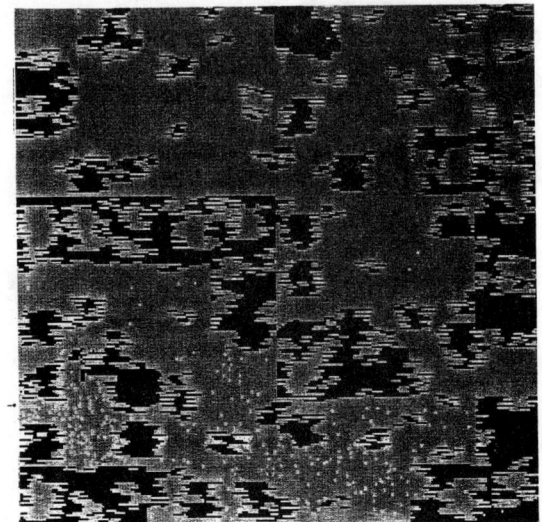

Figure 14.

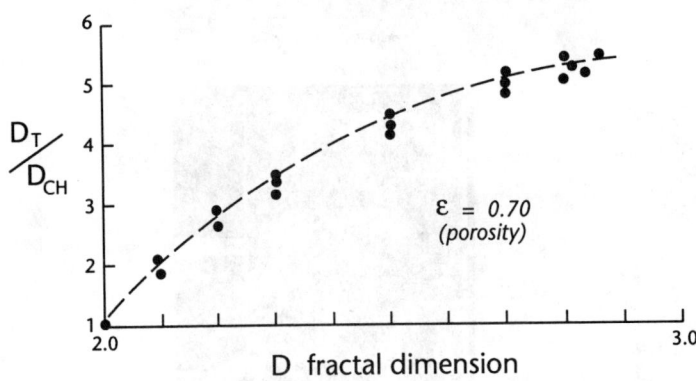

Figure 15. Dependence of normalized transverse dispersion coefficient on fractal dimension.

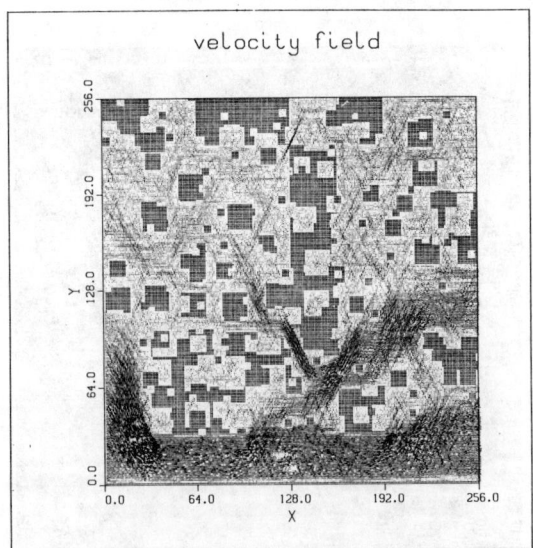

Figure 16.

Understanding the dynamics of transport on the microscale is vital for controlling, e.g., organic contaminant removal and enhanced oil recovery, as well as basic sorption processes. Introducing cohesive forces between particles of different color is the basis for simulating two phase miscible and immiscible flow. Multi-phase flows in porous media via CA's are being studied both by Rothman's group at MIT and at Los Alamos National Laboratory in conjunction with several oil companies.

Fracture Flow

Fractures are a common feature in rocks. Unfortunately, their presence is a difficult challenge for numerical modeling. Models that include fractures only indirectly through averaging processes might be able to represent bulk flow properties, but they will probably fail when aplied to transport. On the other hand, modeling of fracture networks explicitly requires special numerical treatment which is fequently not as robust or computationally efficient as one would wish. For cellular automata, fractures present no special difficulty but can be treated as easily as the surrounding porous matrix.

As an example, Fig. 16 illustrates a fractal porous medium with a fracture running along its base. Velocity vectors are shown. Flow is quite strong in the model fracture, but flow rates in the surrounding matrix are highly variable, from almost stagnant to fairly large in a tributary branching off the main fracture.

These fracture models can be used in several ways. One of importance to underground storage of radioactive waste and to movement of organic contaminants is the calculation of effective diffusion/dispersion coefficients through a permeable formation containing fractures. Diffusion out of fractures can greatly retard the movement of solutes.

Future Directions

A 3-D version of this cellular automata model has been developed and will be used to calculate permeabilities of core samples. Pore structure of small specimens (on the order of a few cms in size) can be determined now with x-ray tomography (Naryanan and Deans, 1988). That structure can then be digitized and used directly in 3-D simulations.

Many applications of this model are possible beyond calculation of permeability and dispersivity. Processes such as enhanced oil recovery and monitoring and control of organic waste sites depend on constitutive relations which successfully scale up from the pore scale. In addition, flow of chemically reactive fluids which modify the porosity and permeability of the host medium can also be treated (e.g., Wells, Janecky and Travis, 1990). Here, CA's can be used to keep separate those effects due to chemical interactions from those due to pore geometry and topology.

Cellular Automata methods allow us to solve the Navier-Stokes equations in arbitrarily complex geometries. The simultaneous development of the etched glass porous media analogues has put us in the fortuitous position of being able to run experiments and perform simulations using exactly the same geometries. Now the effects of pore structure can be accurately determined and separated from chemical/physical effects.

References

Bahralolom, I., Bretz, and F, M. Orr, Jr.,"Experimental Investigation of the Interaction of Phase Behavior With Microscopic Heterogeneity in a CO2 Flood", Soc. Pet. Eng. Reservoir Engineering (May, 1988), 662-672.

Bear, J., Dynamics of Fluids in Porous Media, American Elsevier Publ. Co., New York, 1972.

Chen, S. Y., K. D. Diemer, G. D. Doolen, K. G. Eggert, C. Fu, S. Gutman and B, J, Travis, "Lattice Gas Automata for Flow Through Porous Media", Physica D, (1990), in press.

Dahlburg, J. P., D. Montgomery and G. D. Doolen, "Noise and compressibility in lattice-gas fluids", Physical Review A, 36, 2471-2474 (1987).

Diemer, K., K. Hunt, S. Chen, T. Shimomura and G. Doolen, "Density and Velocity Dependence of Reynolds Number for Several Lattice Gas Methods," in Lattice Gas Methods for Partial Differential Equations, G Doolen, ed., Addison-Wesley, 1990.

Dullien, F. A. L., Porous Media - Fluid Transport and Pore Structure, Academic Press, New York, 1979.

Fournier A., D. Fussell, and L. Carpenter, ``Computer Rendering of Stochastic Models," *Comm. ACM*, **25**, 371-384 (1982).

Frisch, U., B. Hasslacher and Y. Pomeau, ``Lattice-Gas Automata for the Navier-Stokes Equation," *Phys. Rev. Lett.*, **56**, 1505-1507 (1986).

Frisch, U., D. d'Humieres, B. Hasslacher, P. Lallemand, Y. Pomeau and J. Rivet, ``Lattice Gas Hydrodynamics in Two- and Three Dimensions," *Complex Systems* **1**, 649-707 (1987).

Hasslacher, B., ``Discrete Fluids," *Los Alamos Science* No. **15**, 175-217 (1987).

Krohn, C. E., ``Sandstone Fractal and Euclidean Pore Volume Distributions," *J. Geophys. Res.*, **93**, 3286-3296 (1988a).

Krohn, C. E., ``Fractal Measurements of Sandstones, Shales, and Carbonates," *J. Geophys. Res.*, **93**, 3297-3305 (1988b).

Narayanan, K., and H. A. Deans, "A Flow Model Based on the Structure of Heterogeneous Porous Media",Soc. Petrol. Engrs. J.,SPE paper 18328, 675-690 (1988).

Rothman, D., ``Cellular-Automaton Fluids: A Model for Flow in Porous Media," *Geophysics* , **53**, 509-518 (1988).

Wells, J. T., D. R. Janecky and B. J. Travis, "A Lattice Gas Automata Model for Heterogeneous Chemical Reactions at Mineral Surfaces and in Pore Networks", Physica D Proceedings of NATO Conference, (1990), in press.

Wolfram, S., ``Cellular Automaton Fluids 1: Basic Theory," *Journal of   Statistical Physics* , **45**, 471-526 (1986).